INTERNATIONAL ENERGY AGENCY

COAL PROSPECTS AND POLICIES IN IEA COUNTRIES

1981 REVIEW

INTERNATIONAL ENERGY AGENCY

2, RUE ANDRÉ-PASCAL 75775 PARIS CEDEX 16, FRANCE

The International Energy Agency (IEA) is an autonomous body which was established in November 1974 within the framework of the Organisation for Economic Co-operation and Development (OECD) to implement an International Energy Program.

It carries out a comprehensive programme of energy co-operation among twenty-one* of the OECD's twenty-four Member countries. The basic aims of IEA are:

i) co-operation among IEA Participating Countries to reduce excessive dependence on oil through energy conservation, development of alternative energy sources and energy research and development;

ii) an information system on the international oil market as well as consultation with oil com panies;

iii) co-operation with oil producing and other oil consuming countries with a view to developing a stable international energy trade as well as the rational management and use of world energy resources in the interest of all countries;

iv) a plan to prepare Participating Countries against the risk of a major disruption of oil supplies and to share available oil in the event of an emergency.

* *IEA Member countries: Australia, Austria, Belgium, Canada, Denmark, Germany, Greece, Ireland, Italy, Japan, Luxembourg, Netherlands, New Zealand, Norway, Portugal, Spain, Sweden, Switzerland, Turkey, United Kingdom, United States.*

The Organisation for Economic Co-operation and Development (OECD) was set up under a Convention signed in Paris on 14th December 1960, which provides that the OECD shall promote policies designed:

- to achieve the highest sustainable economic growth and employment and rising standard of living in Member countries, while maintaining financial stability, and thus to contribute to the development of the world economy;
- to contribute to sound economic expansion in Member as well as non-member countries in the process of economic development;
- to contribute to the expansion of world trade on a multilateral, non-discriminatory basis in accordance with international obligations.

The Members of OECD are Australia, Austria, Belgium, Canada, Denmark, Finland, France, the Federal Republic of Germany, Greece, Iceland, Ireland, Italy, Japan, Luxembourg, the Netherlands, New Zealand, Norway, Portugal, Spain, Sweden, Switzerland, Turkey, the United Kingdom and the United States.

Publié en français sous le titre :

POLITIQUES ET PERSPECTIVES
CHARBONNIERES
DES PAYS MEMBRES DE L'AIE

EXAMEN 1981

CONTENTS

I EXECUTIVE SUMMARY

In May 1979, IEA Ministers adopted Principles for IEA Action on Coal and decided that the Governing Board would periodically review "the extent to which the main elements of national coal policies required to provide a cooperative framework within the IEA for expansion of coal use and trade have been developed and implemented". This paper reviews coal prospects and policies in IEA countries, based on the conclusions of the first Coal Review, which took place on 28-29 January 1982. The IEA Coal Industry Advisory Board provided valuable advice and comment throughout the Review. At its meeting of April 1982, the Board adopted a position statement on coal prospects which is included in this report as Appendix 1.

Overview of Coal Prospects to 1990

From 1973-80, solid fuels* supplied about 43% of the growth in total energy requirements. This is all the more remarkable since coal use actually declined slightly from 1973-78, before increasing markedly in 1979 and 1980. Results in these two years were strongly influenced by the rapid and substantial increase in oil prices and supported by the emergence of a comprehensive framework of political support. At the international level, this support included:

- the adoption of the Principles for IEA Action on Coal;
- the political recognition, at Ministerial and Summit meetings, of the economic and energy importance of expanding coal use, production and trade;
- the creation of the Coal Industry Advisory Board.

As well, many countries introduced programmes to support coal conversion and in some cases to discourage or ban oil use where coal could serve.

Countries project that coal use will increase by 50% by 1990, providing an *additional* 347 Mtoe of energy by that year. This would be equivalent to the current oil output of Saudi Arabia and account for 50% of the total projected growth in energy requirements between 1980 and 1990. A critical question affecting the overall pace of structural change in IEA countries, and hence the overall vulnerability of IEA economies to medium-term developments in the world oil market, is whether these projections are likely to be achieved.

---

* Solid fuels include coal (including lignite) and other solid fuels such as peat and wood wastes. Available data do not yet allow a consistent breakdown of historical and projected figures. In 1980, however, coal (including lignite) accounted for more than 95% of total solid fuels.

The resource base is not an impediment to expansion of coal use on this scale or in this time frame. The development of associated infrastructure, while sometimes difficult and expensive, can in most cases be completed in time to avoid serious bottlenecks if adequate assurance exists that it will be utilized when completed.

Although the necessity for producers to ensure stable supplies at competitive prices continues to exist, the balance of uncertainties in the current situation are on the demand side, primarily because of:

- the recent weakening of oil prices and increase in coal prices, which have lessened the very strong competitive advantage coal enjoyed and heightened uncertainty about the future evolution of coal prices in relation to competing fuels;
- low economic growth which particularly affects investment in new coal-fired plant and equipment;
- relatively low and highly uncertain projections of electricity demand growth due to continuing low economic growth and improving efficiency; and
- continuing concern about the environmental impact of using coal, the costs of complying with environmental standards, and uncertainty as to whether these standards will change.

On balance, countries' current coal projections are more modest than previous forecasts. They could possibly be achieved but, in the current situation, this is likely to require continued efforts by government — to create a climate conducive to expanded coal use, production and trade — and by industry — to take advantage of the opportunities available to aggressively market coal, to contain pressures leading to higher production or transportation costs, and to achieve economies of scale in maritime transportation.

Coal Demand

Although a considerable increase in coal demand may eventually result from requirements for synthetic fuel production, coal-fired district heating, and even maritime bunker fuel, over the next decade the primary demand for coal in IEA countries will come from the electric power and industrial sectors.

Electricity

Member countries expect the share of coal and other solid fuels to fall from the current share of about 66% of total fuel inputs for electricity generation to about 62% by 1990. Based on countries' current assumption that electricity demand will grow by 3.4% per year from 1980-90, this would require a 46% increase in installed coal-fired capacity and an increase in solid fuel use of almost 179 Mtoe.

Favourable economics compared with oil, suggest that both nuclear power and coal are preferred choices to provide baseload power, with the precise choice between them depending not only on economic factors but also on other concerns such as available sites, regulatory regimes and public attitudes. Coal continues to have a significant economic advantage over oil for new plants, even taking into account environmental costs; but recent price developments have lessened the economic advantage of new coal plants compared to existing oil-fired plants and have adversely affected the economics of converting existing oil-fired plants to coal.

The fact that 50% of the expected additions to coal-fired capacity anticipated by 1990 is not yet under construction, given average lead times of 6-8 years, tends to support the view that current projections may not be realized without more determined efforts by electric utilities and governments. Appropriate actions could include:

— clarifying the desired mix of coal and nuclear power for baseload generation and the prospects for coal to meet medium load requirements.

— encouraging electricity use where it makes economic sense to do so.

— ensuring that regulatory and administrative processes provide a reasonable degree of certainty, so as to avoid unanticipated delays that add to construction costs.

— ensuring that electricity tariff-setting procedures provide adequate revenue to undertake expansion.

Industry

Industry currently uses about 23% of all solid fuels and is expected to be the fastest growing market for coal over the 1980s in relative terms, with total solid fuel use forecast to increase by 76% (133 Mtoe) by 1990. Of this total increase, 85 Mtoe is expected to be accounted for by steam coal.

The cement industry has been the leader in converting to coal, but this industry is uniquely free of some of the difficulties that beset other industries. Major constraints include:

— the lack of available space for coal yards and ash disposal;

— the current economic recession and high interest rates which make new investment difficult;

— the lack of emphasis on reduction of energy costs when these constitute only a small share of total costs;

— difficulties encountered in getting permits to install or convert boilers;

— uncertainty about the reliability and cost of internal distribution systems;

— in some countries, environmental standards which are complex and, at times, more stringent for coal than for oil; and

— in many cases no clear perception of the economic advantage over the long term of using coal rather than oil.

While the competitiveness of coal will help achieve coal conversion, the constraints noted above will tend to slow down otherwise economic decisions. In order to overcome these constraints, IEA countries should pay continuous attention to the following elements:

— lessen the impact of institutional or regulatory barriers that tend to impede the operation of market forces or unduly disadvantage coal in relation to oil;

— keep under review fiscal measures that can disadvantage coal's ability to compete economically with other fuels; and

— consider positive measures that can help to increase information about prospects for greater industrial use of coal and, where appropriate, selective assistance to promote more rapid market penetration.

It is also particularly important for potential industrial users that RD&D efforts be continued in new coal-using technologies, including fluidized bed combustion, coal-liquid mixtures, and coal gasification.

Coal Production

IEA countries expect solid fuel production to increase by about 40% (300 Mtoe) between 1980 and 1990. There are, in general, three groups of coal producing countries in the IEA.

i) High-Cost Producers

Germany, the United Kingdom, Belgium, Spain and Japan are among the coal-producing countries with higher-than-market production costs, at least for bituminous coal. In 1980, these countries produced 195 Mtoe of solid fuels (25% of total IEA production). Production is expected to increase only marginally to 1990, with noticeable gains in Spain and the United Kingdom being partially offset by slight declines in Belgium and Germany.

Production in these countries provides an important base source of energy as well as contributing to other social and economic objectives. It is important, however, that high-cost IEA producers:

— continue efforts to rationalize coal production and increase productivity and strive to eliminate subsidies as quickly as possible where this is feasible;

— not accompany subsidization measures, either overtly or covertly, with quotas, tariffs or administrative arrangements which restrict the ability of imported coal to displace oil.

ii) Countries with Modest Production but Good Potential

Greece, Ireland and Turkey fall into this category. These countries have low Btu coal or lignite reserves, the development of which will be expensive. In these countries, a balance needs to be struck between continuing expansion of domestic production and lower-cost imports.

iii) Low-Cost Major Producers

Australia, Canada and the United States produced 556 Mtoe of solid fuels in 1980 (71% of IEA production). The United States alone produced 471 Mtoe in 1980 and expects production to increase by

38% by 1990. Canada expects production to grow by about 65%, while Australia forecasts an increase of 125% over this decade. All three countries expect to be substantial net exporters by 1990.

Foreign investment in mining activities is unrestricted in the United States, but subject to review and control in both Australia and Canada. The vast majority of applications in these countries has been approved and there are no indications that the review process has constrained the pace of coal production.

Whether production forecasts in these countries materialize will be primarily determined by demand and the capacity of infrastructure to move coal to markets, domestic and foreign. There is a possibility that shortages of skilled manpower may constrain future expansion in Australia and Canada, but governments and industry are working to develop appropriate training programmes. The Australian government is also devoting considerable efforts to improving the overall climate of industrial relations in the coal and related infrastructure industries.

Coal Trade and Infrastructure*

Total IEA steam coal imports are expected to increase from 44 Mtoe in 1980 to 128 Mtoe in 1990. Some of these imports will continue to be supplied by non-IEA countries and some IEA exports will also move outside the IEA area. But, on an overall basis, IEA countries expect to produce about 97% of total IEA requirements in 1990 and the bulk of the growth in IEA steam coal trade will be among IEA countries.

The smooth expansion of trade in steam coal requires, above all, stable and confident long-term relationships among buyers and sellers. Long-term contracts have a key role in developing such relationships and, in particular, in reducing uncertainties associated with the development of infrastructure throughout the coal chain. Other important elements in developing these relationships include:

- the clear formulation by importing countries of longer-term objectives regarding the role of coal and coal imports in their energy systems.
- the clarification by exporting countries of their longer-term objectives with regard to the administrative regime governing export sales and the fiscal treatment of coal production, transport and exports.
- the ability, where appropriate, to build long-term relationships through joint ventures.
- an understanding from governments that commercial trading arrangements, once entered into by buyer and seller, and approved by government where necessary, will not be subsequently interfered with or altered unilaterally.

* An interim report to IEA Ministers of the CIAB Task Force on Infrastructure is attached as Annex 1.

Each of these elements requires serious attention.

Although infrastructure development is not seen to be a major problem, in general terms, there are some specific issues that require continuing consideration. These include:

— policies for dredging access channels on the East and Gulf coasts of the United States, and sharing the costs of dredging, which have not yet been finalized although some positive steps have recently been taken to facilitate dredging at the port of Baltimore.

— the ability, in the United States, of slurry pipelines to compete effectively with other modes of long-distance coal shipment under current institutional arrangements.

— upgrading rail transport capability in Canada.

— the impact of railroad tariffs on coal movement, and particularly the movement of coal for export, in Australia, Canada and the United States.

— the timely development of adequate reception and inland transport facilities in many importing countries, particularly Austria, Belgium, Italy, Greece, Portugal, and Turkey.

Coal Prices and Economics

The expected future evolution of coal prices in relation to oil prices will be fundamental in determining how quickly and to what extent coal displaces oil in IEA energy systems. The very rapid increases in coal prices that occurred over the past two years were in part due to the rapidity with which overall demand grew and the resulting bottlenecks caused by the inability of existing infrastructure to handle such a rapid increase. This general phenomenon was also reinforced by a number of particular events including the disruption of Polish supplies, a strike in Australia and anticipation of a strike in the United States. Recent price increases have nevertheless led to increased uncertainty on the part of many coal users.

Over the longer run, as the volume of trade in steam coal grows and infrastructure is further developed, coal prices should be largely determined by international market forces and should be less affected by constraints such as lack of infrastructure in exporting and importing countries. Production costs will play an important role in determining the international price. The high degree of competition in the United States domestic coal market should exert a restraining influence on coal prices in international trade, since United States' exports are not expected to amount to more than about 15% of domestic production.

It is difficult to say whether coal prices are likely to increase in real terms over the longer term. In many countries the general tendency will be for real costs of production to increase, but scope may well exist to offset this general tendency through greater rates of productivity increase or the achievement of substantial economies of scale in coal transportation, primarily through large vessels. It is equally if not more difficult to predict where real oil prices are likely to go, although all of the long-run scenarios developed by the IEA suggest that basic supply and demand trends point once again to upward pressure on oil prices in the late 1980s and early 1990s.

What is clear is that every effort must be made by those in the coal industry to minimize cost increases and strive for maximum productivity gains in order to protect and enhance coal's competitive position. Governments should also carefully assess the impact of fiscal measures on coal's ability to compete with other fuels.

Environmental Questions

The environmental objections to coal primarily relate to its use, since reclamation regulations have proved workable and affordable at the production end, and transportation impacts are more a nuisance than a threat, and may be reduced by the development of slurry transport technology. Environmentally acceptable coal combustion does involve, however, the control of some significant pollutants which have potentially serious implications, some of which are not yet fully understood. The economic costs of controlling these pollutants can vary considerably from case to case depending upon site-specific factors and differences in mandated levels and types of controls. In general, it appears that sulphur oxides and particulates can now be controlled economically in new large plants, although costs are still a problem in existing plants, and technological developments may be needed to permit economic control in small-scale uses. The situation regarding cost-effective control of nitrogen oxides is less clear.

In general, achieving current environmental standards is not incompatible with increased use of coal. But care will have to be taken by governments, in cooperation with industry, to ensure that regulatory frameworks allow environmental standards to be met at minimum cost. In particular, greater emphasis should be given to:

- prescribing results required rather than means of attainment, thus leaving maximum flexibility to develop appropriate approaches and new technologies;
- ensuring that environmental regulatory procedures do not result in unduly long and uncertain lead times that can unnecessarily increase costs of compliance;
- integrating and coordinating the development of environmental standards and regulations for air, water and land protection, in order to avoid duplication that can lead to higher compliance costs.

II CONCLUSIONS OF THE COAL REVIEW

A. Overview of Coal Prospects to 1990

As IEA countries continue to reduce their dependence on oil, particularly oil imported from politically unstable regions, coal will be among their most important assets. Current projections show that under a continuation of the presently favorable circumstances, solid fuels — overwhelmingly composed of steam and metallurgical coals — will provide 45% of the incremental energy growth expected by IEA countries between 1980 and 1990 (Table 1). Nuclear energy will contribute 33%, natural gas 14%, and hydroelectricity and renewable sources will provide almost equal shares of the remainder. This *additional* amount of coal is equivalent to 347 million tons of crude oil, almost as much as current oil output from Saudi Arabia and equal to the total projected 1990 oil consumption of Australia, Canada, Germany, Ireland, the Netherlands and the United Kingdom.

Table 1 **Solid Fuel* Developments in the IEA (Mtoe)**

	1973	1978	1979	1980	1990
IEA Total Energy Requirements	3335.1	3541.0	3621.1	3514.0	4208.0
IEA Solid Fuels Requirements	690.9	680.1	730.7	767.7	1114.9
Share of Solid Fuels in TPE (%)	20.7	19.2	20.2	21.8	26.5
	1973-78		1978-80		1980-1990
Growth in Total Energy Requirements	205.9		26.9		694.0
Growth in Solid Fuel Requirements	- 10.8		87.6		347.2
Incremental Energy Provided by Solid Fuels (%)	- 5		n.a.		50

Source: OECD Energy Balances and Country Submissions

* Solid fuel includes coal and other solid fuels such as peat and wood wastes. Statistical information does not yet allow a consistent disaggregation of solid fuels. In 1980, however, coal accounted for more than 95% of solid fuel requirements.

Coal, once the dominant fuel in industrial nations, is beginning to reassert itself, driven by the combined forces of favourable market economics and supporting government policy. Despite the oil embargo of 1973, which put the world on notice of the insecurity of oil supplies and which raised oil prices by a factor of four, solid fuel use among IEA countries actually declined absolutely until 1978. It grew rapidly in 1979 and 1980, however, and over the 1973-80 period accounted for 43% of incremental energy requirements in the IEA. The oil price increase in 1979 and 1980 propelled coal into a pattern of rapid growth, and

Member Countries are forecasting that this growth will continue through the 1980s. It will be necessary, however, if these expectations are to be realized, for coal to maintain a clear economic advantage compared to oil and for Member governments to continue to implement policies that support increased coal use, production and trade in order to minimize the security risks of excessive dependence on oil.

Current projections show that coal's contribution to Total Primary Energy (TPE) requirements in IEA countries will rise from one-fifth to one-fourth over the coming decade, while oil use will fall from about half to about three-eighths of TPE*.

On a sectoral basis, this transformation is expected to be concentrated in industrial use and electric generation. Geographically, a substantial increase in coal's share has already occurred in Denmark and is projected to occur in virtually every other IEA country. For example, the share of solid fuels in meeting energy requirements is forecast to treble in the Netherlands, double in Italy, and to almost double in Sweden and Greece. Solid fuels are projected to increase their role in meeting Portugal's energy requirements from 4% in 1980 to 23% in 1990. (Table 2)

Table 2 **Solid Fuel Requirements**

	1973		1980		1990	
	Mtoe	Share of TPE (%)	Mtoe	Share of TPE (%)	Mtoe	Share of TPE (%)
Canada	15.05	8	26.15	11	36.50	14
United States	360.04	20	405.00	22	565.00	28
NORTH AMERICA	375.09	19	431.15	21	601.50	26
Australia	24.97	43	32.68	43	53.40	49
Japan	59.80	18	66.43	18	111.80	22 [1]
New Zealand	1.63	17	1.73	15	2.51	15
PACIFIC	86.40	21	100.84	22	167.71	23
Austria	4.26	18	4.38	16	5.05	14
Belgium	11.72	25	11.58	25	15.00	26
Denmark	2.06	11	6.14	32	8.20	41
Germany	83.96	32	82.70	30	95.00	31
Greece	2.28	19	3.52	22	10.40	40
Ireland	1.58	21	1.89	22	5.04	39
Italy	10.30	8	13.24	9	34.00	19
Luxembourg	2.40	50	1.81	50	2.20	47
Netherlands	3.23	5	4.05	6	12.60	18
Norway	0.76	4	1.44	6	1.60	6
Portugal	0.59	8	0.46	4	5.05	23
Spain	10.07	18	15.31	20	24.26	23
Sweden	4.77	10	4.87	10	11.27	19
Switzerland	0.57	2	0.57	2	0.80	3
Turkey	11.85	48	13.26	42	41.70	55
United Kingdom	79.04	35	69.62	35	73.50	31
EUROPE	229.43	23	235.64	23	345.67	28
IEA TOTAL	690.92	21	767.63	22	1114.88	27

Source: OECD Energy Balances and Country Submissions
1. Japan forecasts are Secretariat estimates.

* The approach and assumptions which have yielded these projections are presented in Annex 1, along with a more complete presentation of the results.

Coal required by the IEA countries will largely come from IEA countries themselves, with anticipated IEA requirements of 1115 Mtoe in 1990 marginally exceeding expected indigenous IEA production of 1079 Mtoe. (Table 3). Although the IEA will be largely self-sufficient in coal in 1990, some IEA production will be exported to non-IEA countries and some IEA countries will rely to a significant degree on non-IEA exporters for their coal supplies.

Table 3 **Solid Fuel Production**

	1973		1980		1990	
	Mtoe	Share of Solid Fuel Req. (%)	Mtoe	Share of Solid Fuel Req. (%)	Mtoe	Share of Solid Fuel Req. (%)
Canada	11.53	77	25.57	98	41.90	115
United States	380.53	106	471.11	116	649.00	115
NORTH AMERICA	392.06	105	496.68	115	690.90	115
Australia	43.23	173	58.95	180	131.60	246
Japan	18.75	31	13.24	20	13.2	12[1]
New Zealand	1.66	102	1.86	108	2.86	114
PACIFIC	63.64	74	74.05	73	147.66	88
Austria	1.31	31	1.50	34	1.00	20
Belgium	6.19	53	4.44	38	4.20	28
Denmark	—	—	0.26	4	0.30	4
Germany	92.84	111	90.32	108	89.00	94
Greece	1.81	79	3.20	91	8.80	85
Ireland	1.06	67	1.08	57	1.73	34
Italy	1.84	18	1.31	10	2.00	6
Luxembourg	—	—	—	—	—	—
Netherlands	1.28	40	—	—	—	—
Norway	0.29	38	0.58	40	0.60	38
Portugal	0.15	25	0.12	26	1.19	24
Spain	7.18	71	12.50	82	14.76	61
Sweden	3.22	68	3.23	66	7.14	63
Switzerland	0.24	42	0.23	40	0.40	50
Turkey	11.89	100	13.31	100	32.11	77
United Kingdom	78.54	99	74.84	107	77.00	105
EUROPE	207.85	91	206.92	88	240.23	70
IEA TOTAL	663.55	96	777.65	101	1078.79	97

Source: OECD Energy Balances and Country Submissions
1. Japan forecasts are Secretariat estimates.

IEA production will not, however, be as geographically diverse as demand: the United States and Australia will be the largest producers and the primary contributors to international trade, while Germany, United Kingdom, Canada and Turkey also expect to achieve major levels of coal production, largely for domestic consumption. All but two IEA countries anticipate that they will have solid fuel production of their own in 1990. Between 1980 and 1990, production is forecast to increase in fourteen countries, remain essentially the same in three, and decline marginally only in Austria and Belgium. The United States, which now accounts for 60% of IEA production and 53% of IEA requirements, will according to current projections, contribute about two-thirds of IEA coal production and consume half of total requirements in 1990.

The relative trade position of IEA countries will not vary much between 1980 and 1990 (Table 4). According to current projections, Canada will change from a

Table 4 **Net IEA Coal Trade**

	NET IMPORTS				NET EXPORTS			
	1973	1980	1990		1973	1980	1990	
	(Mtoe)	(Mtoe)	Mtoe	% of Solid Fuel Requirements	(Mtoe)	(Mtoe)	Mtoe	% of Solid Fuel Production
Canada	2.8	0.43	—	—	—	—	5.40	13
United States	—	—	—	—	31.54	53.98	84.00	13
NORTH AMERICA	—	—	—	—	28.74	53.55	89.40	13
Australia	—	—	—	—	18.25	27.86	78.10	59
Japan	42.01	51.76	98.6	88[1]	—	—	—	—
New Zealand	—	—	—	—	0.02	0.08	0.35	14
PACIFIC	23.75	23.82	20.15	12	—	—	—	—
Austria	2.93	2.89	4.05	80	—	—	—	—
Belgium	5.31	7.18	10.80	72	—	—	—	—
Denmark	1.99	6.06	7.90	96	—	—	—	—
Germany	—	—	6.00	6	10.34	6.46	—	—
Greece	0.47	0.40	1.60	15	—	—	—	—
Ireland	0.50	0.80	3.31	66	—	—	—	—
Italy	8.38	12.10	32.00	94	—	—	—	—
Luxembourg	2.41	1.82	2.20	100	—	—	—	—
Netherlands	1.74	4.05	12.60	100	—	—	—	—
Norway	0.46	0.79	1.00	63	—	—	—	—
Portugal	0.32	0.33	3.86	76	—	—	—	—
Spain	2.48	3.81	9.50	39	—	—	—	—
Sweden	1.65	1.64	4.13	37	—	—	—	—
Switzerland	0.22	0.53	0.40	50	—	—	—	—
Turkey	—	—	9.59	23	—	—	—	—
United Kingdom	—	1.33	—	—	0.88	—	3.50	5.5
EUROPE	17.64	37.27	105.44	31	—	—	—	—
IEA TOTAL	12.65	7.54	36.19	3	—	—	—	—

1. Japan forecasts are Secretariat estimates.

net coal importer to a net exporter and Germany will become a net importer; both countries will be largely self-supplied. The net coal imports of Japan, Italy and the Netherlands will increase most sharply, while Spain and Portugal also expect to post large increases in net coal imports. Regionally, the Pacific IEA Member Countries, as a group, project a virtually flat level of net imports, with Australia's additional exports matching Japan's additional imports. North America and Europe, however, show an increase in net exports and net imports, respectively.

This expansion of the role of coal and other solid fuels in IEA countries, important as it would be in reducing oil dependence and safeguarding the economic security of industrialized countries, may not meet the declared goal of the Venice Summit that coal use should double by the early 1990s*. Current expectations have been scaled down to reflect the unanticipated slowdown in economic growth and consequent revisions of future energy and electricity requirements.

Nonetheless, if current projections are achieved, solid fuel use in the IEA in 1990 would exceed 163% of the 1978 levels on an oil equivalent basis, and the development of supporting infrastructure in the 1980s, together with the accumulation of experience with coal use, will allow continued rapid growth in line with demand beyond 1990.

These projected increases will not occur, however, without solutions to a number of important problems that currently hinder coal expansion:

— On the demand side, uncertainty about how quickly coal use will expand has increased because of continued low economic growth which has made investment more difficult and, together with efficiency improvements, has led to reduced projections of future energy and electricity requirements.

— The recent weakening in real oil prices has lessened coal's current competitive advantage and heightened uncertainty about future price trends.

— The resource base is not an impediment. However, infrastructure to produce and transport coal has to be provided in time to prevent bottlenecks from occurring and to overcome the economic and physical problems of greatly expanded coal production use and trade.

— Concern continues to exist about the environmental effects of using coal, although there are some promising technological opportunities presently being researched that could overcome these problems. It will be essential, however, to rationalize the total costs of the coal chain, including costs of environmental compliance, through technological change and institutional improvements, in order to maintain coal's competitive advantage.

— Further efforts will also be required to develop and sustain a climate of confidence for international trade in coal.

A major purpose of the Coal Review has been to improve the flow of information about coal prospects and policies, so that plans all along the coal chain can be made with greater certainty and confidence. As well, in response to a

* On a tonnage basis, due to increasing use of lower-heat-content coals, the target may be more attainable than when it is interpreted in standard energy units, but the data are not available to allow such projections.

recommendation from the Coal Industry Advisory Board, the IEA is currently designing a comprehensive Coal Information System, which will focus on historical developments and future prospects all along the coal chain. The first phase of this information system will be implemented in 1982. The system is expected to provide a source of information to potential investors in coal use, production and trade, and when fully operational, would also provide a useful framework for future reviews of coal prospects and policies in IEA Member countries.

In addition to better information, however, firmer policy efforts will also be required if the problems identified above are to be overcome and current expectations for coal use, production, and trade are to be realized. Producers will have to continue to make efforts to minimize costs and provide stable supplies at competitive prices; users must clarify the role they expect coal to play in their energy mix and take necessary steps, including investment, to bring this about.

B. Coal Demand

*Electricity generation**

Electricity generation now accounts for 66.3% of solid fuel use in the IEA area. It will continue to be the major market for coal, accounting for a projected 61.8% of all solid fuel use in 1990. In absolute terms, solid fuel use for electricity generation is expected to increase by 35.2% (179.1 Mtoe) from 1980-90.

The use of solid fuels in electricity generation varies widely among countries and even though it is expected to increase substantially over the decade in absolute terms, the disparities will remain. Countries where solid fuels account for a major share in electricity generation are the traditional big producers: the United States, Australia, Germany and the United Kingdom. Other countries with substantial solid fuel use for electricity generation are Denmark, which imports all its coal, Belgium, Spain, Greece and Turkey. Coal use for electricity generation is expected to continue to increase substantially in Australia and the United States. Indeed, the forecast growth in United States consumption would account for about 50% of total growth in IEA coal use for electricity generation. Only in the United Kingdom is coal use expected to decline by about 7% over the 1980s, mainly due to the important development foreseen for nuclear power, while oil use is expected to remain at roughly current levels (below 10 Mtoe).

The future development of coal use in electricity generation depends primarily on the growth of electricity demand. Electricity growth rates differ among countries, but in general have been and are expected to continue to be higher than GDP growth rates. For the IEA as a whole average annual electricity growth rates in the periods 1980-85 and 1985-90 are projected to be 4.0% and 3.5% respectively. But the range anticipated over the next decade varies from about 2% in Belgium to 5% in Italy.

Coal-fired electricity generating capacity is forecast to increase by about 50% from 377 GW at the end of 1980 to 577 GW in 1990. Of this increase, however, only about 50% is currently under construction (Table 5).

* Country by country breakdowns and projections of solid fuel use in electricity generation, by quantity and share, are displayed in Annex 2.

Prospects for coal are constrained by substantial uncertainty about future growth rates for electricity, because of relatively low current and prospective economic growth rates, and by the unclear situation in many countries about the future contribution of nuclear power to overall electricity supply. Countries should clarify the role they see coal playing in their total electrical system as precisely as possible, in order to create a stable planning framework in which long-term procurement of coal and construction of necessary transportation and power generation facilities can take place. More detailed and realistic estimates of planned capacity additions, for both coal-fired plant and nuclear generation, would contribute to this process.

Table 5 **Expected Increase in Coal-Fired Electricity Generating Capacity**
(GW) [1]

	Installed (end 1980)	Projected 1990	Expected Increase	Planned Conversions	Under Construction	Net Planned Additions But Not Yet Under Construction [2]
North America	247.0	357.3	110.3	8.0	54.7	47.6
Pacific	20.3	53.9	33.6	0.7	18.6	14.3
Europe	110.4	171.4	61.0	9.0	19.8	32.2
TOTAL	377.7	582.6	204.9	17.7	93.1	94.1

1. Preliminary figures.
2. Planned construction less planned decommissioning.

Favourable economics compared with oil imply the use of both nuclear power and coal to provide base-load electricity generation in virtually all IEA countries. Nuclear is perceived as the major contributor to future base load requirements, assuming current projections are realized. When electric utilities are located where there is good access to economically attractive coal deposits, coal will provide base-load power and can also contribute to the generation of medium-load electricity. Other fuels can provide power for peak loads more flexibly and economically than coal or nuclear. Indicative examples of generating costs in Europe, the United States and Japan *(for oil, coal and nuclear power)* are in Annex 3. These examples suggest that *new* coal-fired power plants may be more economic than continued operation of *existing* oil-fired power plants in some cases, although the situation will of course vary greatly in different countries and locations, and recent developments in oil and coal prices have tended to lessen the overall economic advantage.

The conversion of existing oil-fired power plants is an option which has been pursued more or less vigorously in different IEA countries, due to differing perceptions of the advantages and technical feasibility of burning coal in equipment now using oil. In Germany, due to the age of most monovalent oil-fired power plants, conversion is not seen to be an economic solution. Newer bi- or trivalent units were switched to coal without difficulty, and the operation of older units is generally restricted to peak loads. In other countries, such as Spain, conversion has not yet started, or is at the planning stage (for example, in the Netherlands). Conversion is being actively pursued in Italy and Japan, and has been successfully achieved in Denmark.

The play of market forces and energy prices, together with the policies and attitudes of governments, will determine the adequacy of incentives for such conversions. However, even when significant government incentives are present (as in Japan and and parts of Canada) or are being introduced (as in Italy) the actual degree of conversion will ultimately depend on general economic conditions, the financial capability and capacity positions of the utility industy, and the problems of overcoming technical constraints such as adequacy of space for coal storage, preparation facilities, waste disposal, and access to coal supply. In the case of oil-fired power plants which were designed to use only that fuel, such technical obstacles may render conversion economically unachievable. In such cases, the options are to shift these plants to peak power use or, as is feasible economically in many instances, to build new coal-fired power plants to replace old baseload oil-fired plants. It would be useful to have a broader exchange of information on the factors motivating and constraining individual decisions.

Regulatory practices and rate-setting have, in a few countries, negatively affected the financial health of electricity utilities. This situation has led to difficulties in raising capital for investment in development programmes, particularly in the United States and Italy. Many initiatives are being considered and taken to ease the situation, but this constraint requires continuing attention.

In many countries, long regulatory and administrative processes are required for construction of coal-fired power plants, causing widely-varying delays. In some cases, as in the United States, this could represent an impediment to the construction of some new power plants, which may require as much as 12 to 15 years to bring on line, although the average is 6 to 8 years. Work is underway to streamline the procedures. In Japan and most European countries, it is a standard process to obtain consensus of the local population and approval by government entities before construction begins. As a consequence, lead-times for plant construction are usually much shorter, and costs are similarly reduced.

A major issue for some European countries is the strength of the government and utility commitment to deal with the constraints which hamper coal use in this sector, including the lack of infrastructure required to unload and transport coal; long regulatory and administrative processes required for construction of coal-fired power plants; and increasing environmental concern. This is particularly serious in Italy where oil use for electricity generation is expected to increase over the next five years and still amount to about 24 Mtoe even in 1990.

To achieve their 1990 projections governments and utilities must take decisions promptly on these issues. In some cases, there is a risk that government will not give enough support to coal development programmes with the result that when economic activity picks up, oil may be the only fuel available to meet increased electricity demand.

*Industry**

The industrial market now accounts for 22.8% of solid fuel use (including metallurgical coal). This is expected to be the fastest growing market for coal over this decade, and to account for 27.7% of solid fuel use in 1990. In absolute

* Annex 4 displays detailed projections and data on solid fuel use in industry.

terms, coal use in industry is projected to increase by 76% (133 Mtoe) between 1980 and 1990. Of this increase, 85 Mtoe is expected to be accounted for by steam coal.

The major industrial coal-consuming sector is iron and steel, but here of course, metallurgical grades of coal are required. The iron and steel industry used about 155 million tons of coke in 1979 to produce about 404 million tons of crude steel. In 1980 a decline in crude steel output of about 35 million tons was recorded, with a parallel reduction in coal use of about 25 million tons of coke. The specific consumption of coke per ton of crude steel will continue to decrease due to the modernization of steel producing capacity but, during the current decade, a large part of the oil and gas consumption (about 50 Mtoe in 1979) will gradually be phased out, and this will partly offset the decline in coal use. On balance, a level of 370 million tons per year of crude steel production in the 1980s with an input of solid fuels close to 200 million tons of coal (about 140 Mtoe) could be considered a satisfactory performance considering the major structural difficulties faced by a large number of steel industries in the IEA area.

Among the industries which can consume steam coal, the cement industry has achieved a very substantial degree of conversion to coal. This performance has been possible because of the relatively high share of energy costs in the cement production, the clear economic advantage of coal and the lack of major environmental and technical problems. Complete conversion has been achieved in Denmark and will be realized in 1982 in Germany, the United Kingdom, and Japan. In Western Europe coal accounted for about 45% of total fuel input to this industry in 1980 and projections indicate that by 1983 the coal share will reach about 80%, although in some countries such as Italy, this conversion is proceeding more slowly. In North America, coal is expected to provide 90% of the energy needs in the cement industry by 1990.

Unfortunately other industries are not in as favourable a situation as cement. Major constraints include:

- the lack of available space for coal yards and ash disposal;
- the current economic recession and high interest rates which make new investment difficult;
- the lack of emphasis on reduction of energy costs when these constitute only a small share of total costs;
- difficulties encountered in getting permits to install or convert boilers;
- uncertainty about the reliability and cost of internal distribution systems;
- in some countries, environmental standards which are complex and, at times, more stringent for coal than for oil; and
- in many cases no clear perception of the economic advantage and increased security of supply over the long term of coal use instead of oil.

The CIAB has recently studied the potential for using coal in the industrial sector in some detail, examining both individual sub-sectors as well as, for each sector, the various ways in which coal might be used.* The major conclusion of

* *The Use of Coal in Industry: A Report of the CIAB* (OECD/IEA, Paris: April 1982)

the report is that the potential for increasing coal use in industry is greater than hitherto realised. Technically and economically, a *steam coal* potential of up to 500 Mtce (350 Mtoe) exists in OECD industry up to the year 1990, and 700 Mtce (490 Mtoe) in the year 2000. The study cautions, however, that the diversity of industrial activity is such that a variety of approaches has to be sought to deal with a number of difficult constraints — including industrialists' perception of conversion convenience and the lack of cash resources to make investments — that could easily delay decisions to convert.

At the heart of the problem is the existing stock of boilers and furnaces which turn over only slowly. The decision to convert is a balance between fuel and other operating costs and capital required. The CIAB report concludes that governments have the opportunity to ensure that the potential to use coal in industry is achieved and that some measures to stimulate conversion to coal would be desirable, particularly in the financial and environmental areas, in order to encourage a more rapid response by industrialists to underlying market forces.

In general, competitiveness of coal will help achieve coal conversion where economically possible. But, a payback period of more than two years often deters industrialists from an economically appropriate conversion decision. Given the large variety of specific circumstances confronted by the industrial users, selective aids to industry can promote easier market penetration by coal. Such programmes as grants, subsidies, depreciation allowances, soft loans, and tax advantages have been introduced by Japan, Spain, Sweden and the United Kingdom. Even though it is difficult to measure their effectiveness unambiguously, they have been generally felt to be helpful in encouraging coal use. Such measures should be continued and improved, and considered by other countries, in order to help industrialists shorten the effective payback periods by which they must judge their potential conversion investments and help them to achieve sound long-term solutions.

In some cases the lack of experience in dealing with environmental regulations represents a constraint, particularly for relatively small users. This is an area where better information should be available and governments could help. A balance also needs to be sought between environmental, health and safety implications of using coal and the economic and energy advantages, taking into account the cost of abatement technologies and prospects for cleaner combustion methods. For example, in some countries such as the Netherlands, coal is likely to expand its share significantly only if cleaner combustion techniques become available. Efforts are under way in this area, in order to use different qualities of coal, increase the efficiency of boilers and turbines, reduce or remove the necessity of flue gas desulphurisation (by removing sulphur before or during combustion) and concentrate ashes in centralised disposal yards. Research programmes into new combustion techniques are also under way in Germany, Japan, the United Kingdom and the United States, and these are supported by considerable governmental subsidies and grants. The most promising technology for wide application in the near future is atmospheric fluidised bed combustion (AFBC). Many boilers using this technology are in operation or under construction in Germany, the United Kingdom and Japan where an IEA programme started in 1980. If this technology proves to be effective in the next few years, it can be expected to spread before 1985 and could be a significant factor in promoting greater industrial coal use for boilers.

It could also have some impact in electricity generation and in the residential/commercial sector by facilitating the use of coal in district heating and individual heating plants.

Another matter of concern, related to part of the coal chain not under direct control of producers or consumers, is the set of issues related to movement of coal to the point of consumption. In many cases coal transportation absorbs a substantial share of the price advantage of coal compared to other fuels, reducing its attractiveness to end users, and reducing coal commerce in general. Coal distribution systems should be given greater attention in order to provide better service to consumers, particularly with regard to such factors as regularity of supply, reasonable assurance of price stability, technical assistance during conversion process, and removal of ash.

*Residential/Commercial Sector**

Coal-fired district heating (DH) systems can be a suitable way to allow coal to recapture part of the residential/commercial markets. The capital-intensive character of this system and the need to assure a high utilisation rate of the grid, however, leave only a few countries which can consider the introduction and the expansion of DH. Denmark may be alone in envisaging a large-scale development, with more than one-third of 1985 heating demand supplied by DH based on decentralized schemes and advanced technologies. In Sweden the expansion of DH is considered to be a very important issue in energy policy. In 1990 about 50% of the total demand for heating is planned to be covered by DH. Germany also has two major projects to interconnect DH systems under construction, supplied by industrial excess heat and combined heat and power generation. Other countries do not now envisage major developments in this area.

Combined heat and power (CHP) plants may provide some heat to DH grids, but the tendency is to supply the excess heat to individual users who can take advantage of bulk heat deliveries without requiring substantial distribution grid development. A major study is currently underway in the United Kingdom and results will be known shortly. Ireland also foresees a CHP project in the next few years. In some countries, utilities do not have the right to sell heat in their charter. Where these or other regulatory barriers to the expansion of CHP exist, they should be reviewed.

In many countries, increasing penetration of electricity (which can be provided by coal) in end use, perhaps through wider use of electric heat pumps, seems to be a more realistic option for replacing oil with coal in the residential/commercial sector.

Synthetic Fuels and Other New Technologies

Production of synthetic fuels from coal is expected to be developed to a moderate extent in the 1980s and could begin to have an impact in the 1990s. In 1981, a special OECD/IEA High Level Group for New Energy Technology

* Annex 5 presents data and projections for solid fuel use in the residential/commercial sector, as well as projected electricity use in this sector.

Commercialisation assessed the prospects for technologies currently being pursued and the possible contribution of synthetics by 1990. The High Level Group estimated the possible contribution of synthetic fuels from coal to be 0.4-1.1 million barrels of oil equivalent per day (Mbdoe) by 1990 and from 1.9-4.9 Mbdoe by 2000.* The lower levels would require feedstocks of about 50-80 Mtce by 1990 and 225-400 Mtce by 2000. In its conclusions, the High Level Group indicated its belief that the significant growth of coal conversion industries would come after the expanded conventional use of coal, changes in oil refining output mixes and the development of primary synthetic fuels production such as tar sands and oil shale. The initial pattern of development for the large unit size coal conversion technologies is likely to proceed on a project-by-project basis, each tailored to the specifics of resource and market. Coal-based synthetic fuel projects are vulnerable to the same uncertainties about future prices of oil and other fuels, and to the same general economic conditions, that currently hinder general coal use expansion.

There are two major areas of concern relative to synthetic fuels. The first relates to the voluntary reduction by many governments of their synthetic fuel development programmes. For example, these efforts have been substantially trimmed over the last two years in Germany and the United States. The need to disseminate more information on the economics of coal gasification technologies is the second concern emerging from the review. Countries with relatively small deposits of low-grade coal, very difficult to utilise with conventional burning technologies, are particularly interested. Among them are Greece, Ireland, Portugal and Turkey. Gasification of coal should presently be considered the major potential contribution by coal-based synthetic fuels in the 1990s, though the liquefaction of easily accessible brown coal in Germany should prove economic within a foreseeable timespan.

There is a general need for greater study and understanding of the strategic objectives and prospects for the systematic development of new synthetic fuel technologies. Careful, but determined, efforts are necessary to ensure that the pace of technology development proceeds in a manner that allows commercial production, when overall economic conditions warrant, with minimum delay and at predictable cost. Attention must also be given to timely development of appropriate transportation and use technologies associated with new methods of production.

A great deal of research on new coal combustion technologies is under way: Atmospheric Fluidized Bed Combustion (AFBC), Pressurised Fluidized Bed Combustion (PFBC), Coal-Oil Mixture (COM), Coal-Water Mixture (CWM) and Coal-Alcohol mixture (CAM) are some of the most important technological opportunities. Actual commercialization has started only for the AFBC. The research in the majority of the other technologies has either reached the stage of development or demonstration, but the general opinion prevails that large scale commercialization will not occur until the second half of the 1980s. The need to maintain or increase RD&D by both industry and governments, despite tight budgets and low economic growth, is indispensable if substantial long-term coal development in industry is to materialize.

* *Report of the High Level Group for Energy Technology Commercialisation* (OECD/IEA, Paris: 1981)

Maritime Transport

Coal could have potential for fueling maritime transportation, particularly in shipping of bulk commodities such as coal. Current IEA bunkering requirements amount to about 70 Mtoe (1.4 Mbd) and these are projected to increase to 78 Mtoe by 1990. In the last two years a small number of coal-fired vessels to provide service as part of a long-term coal supply contract have been considered. Traffic in such vessels would initially be limited to regular coal movements, or movements of other minerals in areas where coal is also available, because of the current lack of coal bunkering stations. Longer term options to increase coal use in this activity should be considered.

C. Coal Production

Considerable progress is foreseen in production of solid fuels in IEA countries in the 1980s, with total production increasing by almost 40%, from 1111 Mtce (778 Mtoe) in 1980 to 1541 Mtce (1078 Mtoe) in 1990. A breakdown of production by IEA region is shown below, and country details are given in Annex 6.

Table 6 **Production (Mtoe)**

	1980	1990	Difference 1980-1990	
			Mtoe	% Increase
IEA TOTAL	777.7	1078.8	301.1	38.7
Europe	206.9	240.2	33.3	16.1
North America	496.7	690.9	194.2	39.1
Pacific	74.1	147.7	73.6	99.3

In Europe, most IEA countries expect to maintain their current production level with the exception of Spain, Greece and Turkey: in the latter two production is forecast to almost treble by 1990, making a considerable contribution to their energy requirements. Canada projects more than a 60% increase in production from 1980-90 and will become a net exporter in the 1980s. The United States expects to increase production by 38% and exports by 47%. A substantial increase in production, by more than two-fold, is planned in Australia, which will maintain the position of second largest IEA exporting country, following the United States.

The coal-producing countries of the IEA generally fall into three groups: high-cost IEA producers; IEA countries with modest production but good potential; and low-cost major IEA producers.

High-Cost IEA Producers

Germany, United Kingdom, Belgium, Spain and Japan are among the coal-producing nations which have high production costs, at least for bituminous coal. There is general recognition of the necessity — for social political and regional development reasons — to maintain coal production in these countries through subsidies, since production costs are generally above world market price levels. Many efforts are being made to rationalize coal industries and increase productivity, and significant results have been achieved in the last few years, particularly in the United Kingdom, Belgium, Spain and Japan.

In the United Kingdom, great efforts, including more than five years of heavy investment and an important productivity agreement with the miners, have led to an advanced stage of rationalization in coal production. Most coal output is already competitive with imports in the domestic market, without subsidy. The trend to rationalization has slowed somewhat because of current economic conditions, but the government plans to eliminate, through rationalization and investment, the remaining subsidy as quickly as possible. A critical factor determining the pace at which this can be accomplished is the extent to which industrial coal use can be increased through switching from oil and natural gas.

In Germany coal production is substantially mechanized and has already reached a good level of productivity; further progress in rationalization is constrained by the growing depth of coal mines which in certain cases may reach 2000 metres. It is difficult to foresee an increase in production because of the many problems faced just to keep production at current levels. New capacity will mainly offset exhausted pits.

In general a balance has to be struck between increased coal production and the subsidies governments are prepared to pay. Over the next decade production increases are expected only in the United Kingdom and Spain, where continuing rationalization efforts and the development of new economic mines should make at least a portion of domestic coal competitive with imported coal over the long term.

Only Germany and Spain have coal import quotas in place. These quotas do not presently constrain import levels and, indeed, both Germany and Spain project considerable growth in coal imports over the 1980s. For the latter country, the level of coal imports is evaluated yearly according to total coal requirements. Major IEA coal producers feel, however, that quotas or other measures that could possibly limit access to markets create an element of uncertainty that can affect investment in increased export capacity. No overt or covert limitations, through quotas, tariffs or administrative arrangements, should restrict coal imports where they can effectively displace oil or gas. It would be a positive step for all consuming countries, including high-cost producers, to define in a clear and stable way their policy objectives for indigenous coal production and the long-term evolution of their need for coal imports.

IEA Countries with Modest Production but Good Potential

The IEA countries which fall into this group, mainly Greece, Ireland and Turkey, have low-Btu coal or lignite reserves which are being developed to some extent, generally in difficult conditions due to the marginal economics of such exploitation and lack of capital for investment. The commonly anticipated use for such reserves is in electricity generation by mine-mouth power plants.

It is generally recognized that a more substantial use of coal reserves could be achieved with further development of combustion and synthetic fuel technologies. Greater international collaboration in and exchange of information on ongoing research is required.

Development of imports might, in the end, be more efficient for the overall economy of some of these countries than continuing the expensive expansion of their own production capacity, given the relatively higher costs and competing needs for capital. The need to strike a balance between these two options will require further attention.

Lower-Cost Major Producers in the IEA

The enormous resources available in Australia, Canada and the United States constitute a basic advantage to the IEA for meeting future coal requirements. The goal in these countries is to expand production capacity to meet the increasing demand for coal.

Major expansions are currently underway. From 1980-90 production is expected to increase by 38% (178 Mtoe) in the United States, 130% (75 Mtoe) in Australia, and 59% (16 Mtoe) in Canada. The United States expects to export about 13% of total 1990 production. Australia and Canada expect to export 60% and 37%, respectively, of their 1990 solid fuel production. Market forces will make possible these expansions if satisfactory long-term contractual commitments are made. However, uncertainty about consistent future demand for coal and, in particular, coal exports is the critical issue faced by both governments and industry at present.

The participation of foreign interests in mining activities is unrestricted in the United States, but subject to review and control in both Australia and Canada. In the United States, foreign ownership now accounts for about 3% of current production capacity, but this share is increasing. In Australia the foreign share in mining activities is limited to 50% but this limit can be, and has been, raised in circumstances where it is clear that Australian capital is not available. In Canada, where domestic interests now own about 65% of mining activities, all proposals for foreign participation are subject to review by the Foreign Investment Review Agency. The vast majority of applications for foreign participation in both Canada and Australia have been approved and there are no indications that the review process has constrained the pace of coal production in these countries.

The pace of production in all countries will be primarily determined by demand and the capacity of infrastructure to move coal to markets, domestically and abroad. There is a possibility that in Canada and Australia shortages of skilled manpower may place future constraints on coal production, but governments and industry are working to develop training programmes to prevent this from occurring.

A substantial expansion of production in Canada is planned. This will require the development of new coal mines with massive capital investment, mainly in areas which are far from consuming centres, and the exploitation of sub-bituminous coal reserves with relatively low calorific value. In order to assure projected exports, major capacity expansion will be required in West Coast ports and connecting railroads. Even with government assistance, firm long-term commitments by users are indispensible as producers currently face severe cash flow difficulties and need such commitments to obtain financing for capacity expansion.

Demand uncertainty in the domestic market is the major constraint facing coal production in the United States. There is currently estimated to be as much as 200 million tonnes of annual surplus production capacity. The environmental problems which once plagued coal production have become much better managed in recent years as effective regulations are being administered with more sensitivity to producer requirements. The removal of the coal-leasing moratorium is also an encouraging development. United States production is not, however, suffering importantly from resource limitations. More critical have been several constraints to coal demand which prevent productive capacity from being fully utilized. Relatively low natural gas prices, for example, constrain increased coal use and a more favourable climate for coal will emerge with deregulation of gas prices. A further major constraint is the environmental standards applying to coal combustion, which require the same treatment standards for all coals, regardless of sulphur content.

In Australia, increased coal production is being strongly pursued. The most recent budget, however, reduced somewhat the very generous accelerated write-off provisions for a wide range of capital expenditure that were introduced in 1976-77. Projections for 1990 show that exports will be higher than domestic consumption, but this will require firm commitments by buyers and shippers in the next few years in order to have new mines and infrastructure in place by that time. This development programme will also require resolution of labour relations and training issues. Both Commonwealth and State governments have recently agreed to deal with labour market supply problems, in close consultation with industry.

Non-IEA Coal Producing Countries

Future coal requirements of some IEA countries will continue to be met by non-IEA producers. A brief review of the situation in some of these countries is provided below. More detailed information on Poland and the Republic of South Africa is presented in Section III.

In the **Republic of South Africa,** for economic and political reasons, coal never lost its prominent role in covering the country's primary energy requirements. Currently coal accounts for about 75% of TPE and the authorities plan to at least maintain this share for the coming decades. Electricity generation and conversion to gaseous and liquid fuels absorbed about 75% of total domestic coal consumption in 1980. Production reached 113 million tons in 1980 and the anticipated level for 1990 is expected to be in the range of 200-240 million tons. Exports reached a record 28.5 million tons in 1980, representing near saturation of current capacity. Development is under way to bring export levels to 44 million tons by 1985-86. Plans to further expand coal exports were announced in September 1981 and exports are now projected to increase gradually to a level of 80 million tons/year for a period of 30 years.

The development by government and industry of port facilities (Richards Bay) and associated railway connections have made South Africa a major actor in world coal trade, and the major non-IEA coal exporter. Coal exports are controlled through export licenses which are, however, granted for long time spans (20 to 30 years). South Africa has large reserves of steam coal and a good record of capacity expansion. A recent reassessment of coal resources has resulted in a doubling of known coal deposits. The adequacy of manpower and capability to achieve timely equipment availability are undisputed, and South Africa has a proven track-record for reliable deliveries of competitively-priced coal. Nonetheless, a degree of uncertainty remains about South Africa's future role in the world market, primarily for political reasons.

Poland regularly increased its coal production from the mid-1950s and passed the 100 million ton mark in 1960 (104 million tons). From 1960 to 1979 production increased at an annual growth rate of about 3.5% to reach 201 million tons and in the 1970s the growth rate averaged 4.1% per year. Since 1980, fundamental changes have affected conditions in the Polish coal industry. Labour unrest and the introduction of the five-day week brought production back to 163 million tons in 1981. Currently the production level seems to be recovering to some extent but it is anticipated that the pre-crisis level of more than 200 million tons per year will not be regained for several years. For 1982, the anticipated production level has been set at 185 million tons. For the next five years it is doubtful that the coal mining industry will be able to make up a great part of its backlog; moreover investment programmes for new capacity have been stopped. Since December 1981 the six-day week has again been introduced but work on Sunday will not apply as was the case before 1980. Lack of foreign currency has impeded adequate maintenance of mining equipment and expected progress in mechanisation, resulting in productivity declines.

The labor unrest and accomanpying events are having severe effects on export capability which has been revised downward from its pre-1980 level of 40-42 million tons/year to 18-20 million tons; in 1981 the export level fell to 15 million tons. Poland has lost its place as the world's second largest exporter. A likely target for total exports could be 15 to 20 million tons depending on how the government strikes the balance between priority requirements of the domestic economy and the need to obtain foreign currency. An acute need for foreign currency may from time to time lead to increases in exports above these levels but the likely duration of such increases should be considered very cautiously. Domestic demand is growing very rapidly with an electricity demand growth rate of between 7% and 10%. This cannot be met at current production levels, resulting in rationing and frequent brownouts.

The USSR, the third largest hard coal producer in the world, (552 million tons hard coal and 164 million tons lignite in 1980) has seen its production for the past decade increase at a relatively slow rate (1.5% per year) which has masked a constant drop in productivity over the last 4-5 years. Depletion of the reserves in the European part of the USSR and an accumulation of constraints in developing new resources located in Siberia are the major reasons for this slow development. The most recent government objectives set the production level to increase by 7.5-11.7% between 1980 and 1985, which would bring total production of solid fuels (hard coal and lignite) to a range of 770-800 million tons. If this can be achieved, 1990 production levels might reach 850 million tons. It is likely, however, that these objectives will not be achieved given the recent unsatisfactory performance of the coal industry.

Compared to total production, export figures are marginal (22 million tons in 1980) but they are crucial for covering hard coal requirements of East European countries such as Bulgaria (100%), Democratic Republic of Germany (60%), Czechoslovakia (50%), and Hungary (about 20%). The export level has been reduced slightly from the 25 million tons/year which prevailed for about five years and it is expected that it will not be higher than 21-22 million tons for the 1981-85 period. There are currently no figures available beyond that time horizon. Although firm information is unavailable, exports from the USSR appear to have fallen recently, possibly due to the need to meet domestic requirements normally supplied by imports from Poland, as well as to substitute Soviet coal for lost Polish exports in other East European countries. In addition to the disruption of Polish coal supplies, the interruption of gas deliveries from Iran to the USSR may also have created additional difficulties in meeting Soviet energy requirements, requiring additional amounts of coal otherwise intended for export.

China is the world's second largest coal producer. During the last decade, production increased by 6.5% per year from 1970-79, reaching a level of 635 million tons in 1979 before falling in 1980 to 600 million tons. The forecast for 1981 production anticipates a level slightly above 600 million tons. Resources are considerable but so is the need to improve infrastructure. After a period of concentrated buying of equipment from Western industrialized countries, Chinese authorities now seem, within their overall economic readjustment plan, to be reexamining some of the most important development projects in order to keep borrowing within manageable limits. This reexamination may affect some large new mining developments with the accent currently being put on the mechanisation of existing collieries, although a feasability study for a joint venture with Occidental Petroleum to develop what would be the largest coal mine in the world (14 million tonnes per year) was announced in March of 1982. The current energy restrictions, in particular the shortage in electricity supply, will require a major effort to increase electricity generation. This will absorb growing quantities of coal.

China is recognized to have a large coal potential which could make available significant quantities for export, but major problems include renovation and expansion of mines, inland transportation, and port infrastructure. It is also not yet clear whether coal or oil will be primarily targetted for export. As a result, there is a great deal of uncertainty about the extent to which China will be in a position to make a contribution to IEA coal supplies. Currently, exports remain modest compared to production and go mainly to Japan. Negotiations are in progress between Japan and China in the direction that the latter will export 10 million tons of coal to the former in 1985. It is unclear how the recently announced development will affect future exports from other areas to Japan.

Colombia should become an exporter of coal by the mid-1980s. A long-term contract for 2 million tons per year has already been signed by a European customer, and production from a new Atlantic coast coal field should begin by 1985. Colombia has increased its coal export target from 15 to 50-60 million tons per year by the turn of the century.

Some other countries, not yet recognised as important producers, have potential to enter the international coal market if both technical and political problems are solved. Coal exploitation in **Botswana** would need the construction of a 1500 km railway and the upgrading of the port of Walvis Bay (Namibia); exploitation of coal in **Swaziland** would require the upgrading of the existing railway to Maputo and the development of port facilities. **India** is trying to develop an ambitious coal programme aiming at a production level of 130 million tons by the mid-1980s but for the last four years the production has been stagnating at around 105 million tons/year. The chances that India will become a major exporter during this decade are scant because of the large domestic needs of the Indian economy.

D. Coal Trade and Infrastructure

Recent Developments

Major developments in world coal trade from 1978 to 1980 are summarised in Table 7. Overall world trade of 250 million tons in 1980 is broken down by origin and destination in Figure 1. Divided by quality, coking coal trade can be estimated at 140 million tons, leaving a growing share to steam coal (110 million tons).

Table 7 **Major Coal Exporters, 1978-80**
(million metric tons of coal)

	1978	1979	1980	1979/80 (%)	1980/79 (%)	1980 Share of Coal Trade (%)
Australia	38.7	40.3	43.0	4.1	6.7	17.2
Canada	13.7	13.9	15.0		7.9	6.0
United States	37.0	58.8	81.6	58.9	38.8	32.6
Poland	40.1	42.0	32.0		- 23.8	12.8
South Africa	15.3	23.4	29.0	52.9	23.9	11.6
USSR	24.8	23.9	23.0			9.2
Total	169.6	202.7	223.6	19.5	10.3	89.4
total World Exports	197.9	229.0	250.0	15.7	9.1	100%

In 1980, several important developments took place in world trade. A decline of 24% in Polish exports occurred, and deliveries were irregular as a consequence of reduced production and increased priority to supply the domestic market. This change will affect traditional importers (Italy, Denmark, Germany, Spain, Austria) for an undetermined period of time, even though long-term contracts have not been cancelled but only reduced for "force majeure" reasons. During 1981 the situation deteriorated further and the expected level of total Polish exports for the current year may be as low as 15 million tons compared to previous levels of 40-42 million tons per year.

In the Republic of South Africa, the 1980 level of exports reached a record 29 million tons, almost 24% more than 1979.

In Australia, several strikes at mines and ports affected the normal flow of exports for part of the year. Nevertheless, coal exports still increased by about

Figure 1

WORLD COAL EXPORTS AND IMPORTS BY COUNTRIES, 1980
(Million Metric Tons)

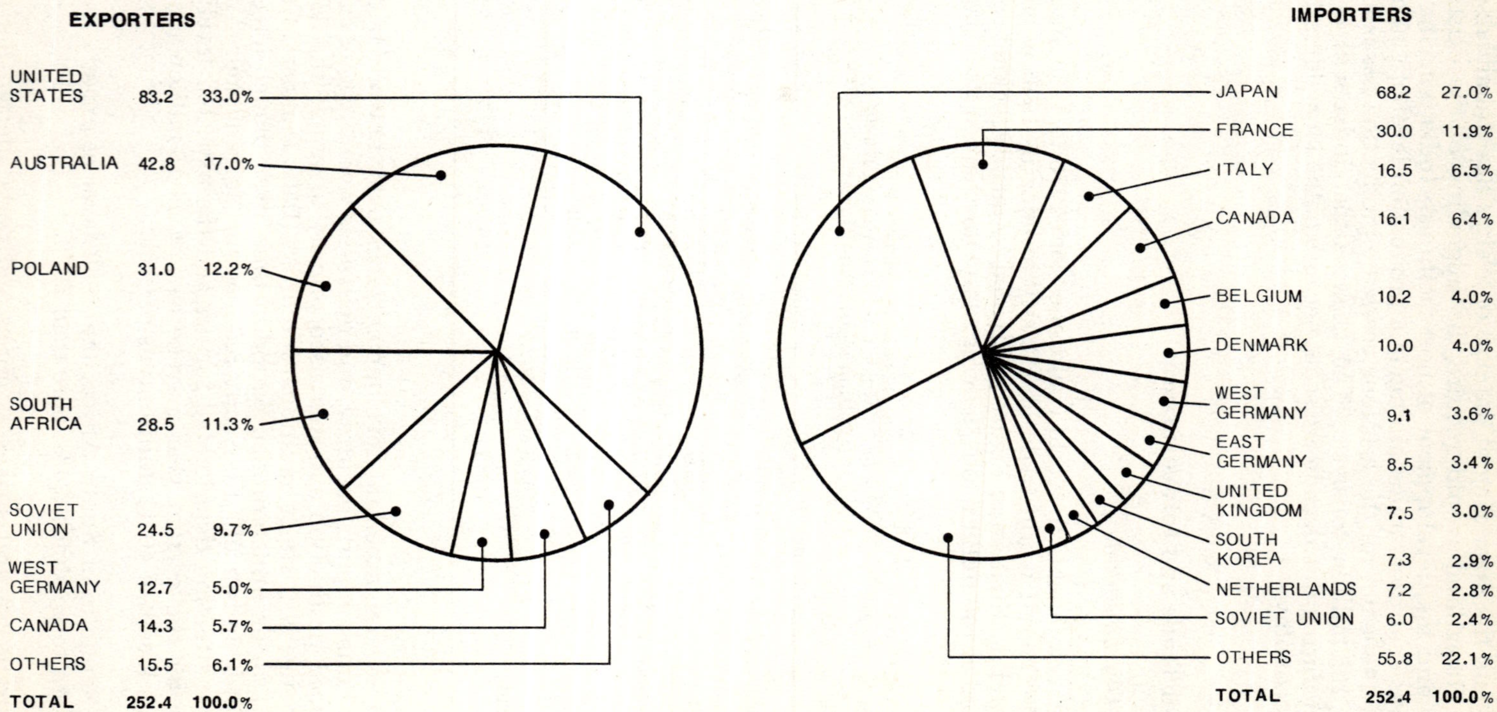

Source : World Coal, November / December 1981.

7%. The Commonwealth government continued to maintain surveillance of coal contracting in order to satisfy itself that contracts reflected fair and reasonable terms. In the United States, exports grew by almost 40%, following a 59% increase in 1979, despite the emergence of bottlenecks in East Coast ports.

Future Prospects

International trade of steam coal is a relative novelty in the international trade of bulk commodities, since for many years only metallurgical coal was traded regularly. Although metallurgical coal trade is expected to increase in absolute terms over the 1980s, sluggish expansion of the world steel industry is producing a decline in its overall share of coal trade at the same time as the attractiveness of coal for energy purposes is increasing. Future growth in coal trade will, therefore, be dominated by the expansion in steam coal.

Coal trade projections of IEA countries are presented in Annex 7. They suggest that coal exports will increase by about 65% from 1980 to 1990 and the largest share of the growth (81%) will be made up of steam coal exports. Coal imports will almost double in the same period and here also steam coal will be the major contributor, accounting for 74% of the increase between 1980 and 1990. IEA countries as a group will tend to become less dependent on imports from non-IEA countries, with overall net imports from outside the IEA amounting to 35 Mtoe in 1990; this requirement, transformed into coal figures, could be as high as 50 Mtce or 56 million tons of coal (at 6300 kcal/kg).

On the exporting side, Canada, the United States and Australia project the biggest increase in the IEA area, and their share of IEA exports would increase from about 80% in 1980 to 94% in 1990. On the import side, rapid growth is expected in a number of countries including Denmark, Greece, Ireland, Italy, Japan, the Netherlands, Portugal, Spain, Sweden and Turkey. In terms of absolute volumes, however, the projections show the greatest growth in Italy, Japan, the Netherlands and Turkey which together would account for two-thirds of IEA total imports in 1990, compared with about 58% in 1980. The overall IEA situation is summarised in the following table.

Table 8 **Projected Import Requirements and Availability**

	1990	
	Mtoe	Million tons of Coal [1]
Total IEA Requirements	1115	1770
Anticipated IEA Production	1079	1713
Anticipated IEA Net Imports	36	57
Potential Availability from Non-IEA countries	62	100
Of which:		
South Africa	38	60
Poland	9	15
USSR	3	5
Colombia	9	15
Others	3	5

1. Coal at 6300 kcal/kg.

About 60% of the coal that enters world trade originates from three IEA countries: United States, Australia, and Canada. The resource base of these major suppliers is adequate to satisfy greatly increasing trade flows, but each of them has specific problems and advantages in relation to particular potential customers.

The United States demonstrated its capability to respond to export demand in a short period of time when exports increased by more than one-third in 1980. But while proving the ability of production capacity to respond, this experience also underscored certain weaknesses, particularly in transportation and port infrastructure. The need for modernization and expansion of port facilities was demonstrated, and projects are now under way which may even lead to some overcapacity in the 1980s, reversing the present situation. But, as discussed below, some important decisions that will affect the long-term cost of coal exports have not yet been taken.

The main advantages for Australia are the dedication of the coal industry to international commerce, and the physical proximity to Far East markets. The Commonwealth government seeks to ensure that exports of coal take place under terms and conditions in line with the world market and to this end approval is required for the export of coal. This power is not used to interfere with long-term contracts for international trade in coal once they have come into force. The Australian state governments are also heavily involved in the approval, development, and operation of export facilities, including the dredging of harbours to allow access of large bulk carriers for more economic transportation to the consumer. Difficult labour/management relations, with numerous strikes of relatively short duration, have affected various parts of the coal chain and resulted in concerns to customers. These have been given considerable attention by Australian authorities, and efforts are being made to improve industrial relations.

Canada is also well located to take advantage of the anticipated growth in steam coal demand in the Pacific area, and should become the third largest IEA exporter by 1990. Coal production in Western Canada (British Columbia and Alberta) suffers, however, from a transportation cost disadvantage to European markets. Moreover, massive initial infrastructure investments, particularly for rail transportation, are required if coal is to be moved to export markets.

The reliance on any one of the traditional major international coal suppliers requires a balance among these problems and advantages which will be found through the negotiation of contracts in the marketplace. In general, however, importers will continue to assure their coal procurement according to a risk management strategy that will entail diversification among suppliers.

Developing Long-Term Trade Relationships

Expansion of steam coal trade requires the development of stable trading relationships among all participants in the coal chain, from the producer to the final consumer. A major difficulty in providing smooth development of this trade lies with the number of participants; the failure of any one of these operators to meet requirements may disrupt the whole process.

Although all major coal-producing IEA countries consider long-term contracts to be an important component of expanded coal trade, there are certain constraints to the actual functioning of long-term contracts. There are many different types of long-term contracts, particularly with regard to variations in price escalation provisions and procedures for renegotiation. Practices vary greatly and some contracts, although purportedly long-term, are so hedged that they do not provide assurances adequate to undertake with confidence the investment required for development of new capacity or infrastructure. In a dynamic market situation there can be no single, or even preferred way, of writing a long-term contract. Overall, however, such contracts should provide an equitable balance of risk between buyer and seller and, to the degree possible, serve to create a commitment that could allow the planning and investment necessary by both parties to proceed with greater confidence.

More fundamentally, there are a number of reasons why, in the current market situation, substantial uncertainty exists among buyers and sellers of coal:

— Major potential importing countries are facing low rates of economic growth. The result has been diminishing profits for both the electric utilities and industries which are the potential users of steam coal. The current high level of interest rates also poses a major constraint to the construction of new coal-fired installations. These factors, together with the increase in coal prices over the last few years, the decline in oil prices registered in 1981 and early 1982 and the continuing weakness in the oil market, have led to a situation where importers appear to be increasingly reluctant to enter into long-term contractual arrangements.

— As well, consumers who are converting to coal, or deciding to use coal in a new use, generally prefer in the first instance to gain experience with an energy source new to them in a measured way. For this reason, they tend to prefer initially shorter-term contracts which allow them to test the market while assessing long-term trends and their energy requirements. After these new coal users have gained sufficient experience, they may well decide to enter into longer term contracts.

— The attractiveness of long-term contracts to importers is diminished to the extent supplies under existing contracts are disrupted by strikes.

In general, the time required to construct new infrastructure to serve a new demand for coal is less than the time required to bring that demand itself on stream. Because of this, it is critically important that appropriate long-term commitments, whether through contractual arrangements, joint ventures or other means, are put in place in time to promote the development of new capacity and associated infrastructure.

Another issue bearing upon the future development of long-term trading relationships has to do with discriminatory fiscal or pricing policies or transportation tariff structures that tend to penalize coal exports at the expense of coal consumed domestically. Examples cited were the Australian levy on steam and coking coal exports, and the higher rail tariffs in the United States for coal destined for export markets. Such practices, as well as providing additional costs, can increase uncertainty and inhibit the development of long-term coal trade.

Governments of exporting countries should clarify their long-term objectives with regard to the fiscal treatment of production and trade, and should carefully assess the impact of discriminatory measures on the attractiveness of continuing long-term arrangements, if not on the very perception of security of supply and price advantage which constitute the major motivations for expanding coal use.

Governments of importing countries should equally clarify their long-term objectives regarding the use of coal in their energy systems and measures they propose to achieve these objectives, in order to provide greater certainty to their own investors and those required to develop new capacity and infrastructure in other countries.

Infrastructure in Exporting Countries

Infrastructure in coal exporting countries, despite good progress, continues to require government attention and encouragement. Several projects are under way in Australia, Canada (West Coast) and the United States (East Coast and Gulf Coast). These are sponsored by governments in Australia and Canada, but in the United States, the initiative lies with the private sector for expansion of port capacities and storage areas.

The issue of dredging in the United States has not yet been resolved. Allowing the loading of larger vessels in selected United States ports holds the promise of reducing substantially the cost of ocean freight. Dredging port facilities on the Gulf Coast is also necessary if the effect of securing cheaper delivery of coal by barge is not to be offset by the increased distance of maritime transport to the consumer. Various proposals regarding the dredging issue are currently being considered by both the House of Representatives and the Senate; these bills contain various ways to settle the financing of these operations which range from total coverage by federal funds to no financial involvement at all with the authorisation to raise a users fee. It is too early yet to give a precise time horizon when a bill might be finalized. A positive evolution has, however, been recorded for the port of Baltimore which received in January 1982, agreement from the federal government to deepen its access to 50 feet which will allow coal carriers of 100,000-110,000 dwt to be fully laden. Final agreement on Baltimore dredging depends on signature of formal contractual agreements by federal and local authorities.

Progress is being made in allowing large bulk carriers (100,000-120,000 dwt) to call at Australian and Canadian ports. A swift solution in that direction for a selected number of United States ports would benefit all coal trade as a better development and use of the world coal carrier fleet could be planned. Maintaining the current situation of limited access to United States ports (PANAMAX-type vessels) could result in artificially distorted fleet development with higher long-run costs. The international coal trade will not benefit fully from economies of scale unless large vessels are able to run indifferently on various routes.

The allocation of the costs and benefits of this development clearly merits analysis and further discussion, but the consensus is that all parties to the trade could be net beneficiaries over the long term from such actions. Action by governments might be needed to strike the balance between diverging regional interests and select among competing ports or different sites which offer about similar basic requisites for subsequent expansion.

The upgrading of railways and the especially difficult matter of railway tariffs also requires further examination. Coal should not lose its competitiveness because of inefficient tariff regulations which penalize the movement of coal in order to provide subsidies to other commodities, or because of the ability of railways to exercise monopoly power (in some cases with State government support) to extract abnormal profits from the coal trade. This is now the case, to greater or lesser degree, in Australia (New South Wales and Queensland), Western Canada and the United States (where coal for export is charged a rate different from coal for domestic use). Railway transport is the segment in the coal chain which is at the same time most limited in adequacy and least susceptible to improvement and control by market forces alone. Government action should aim at putting the railway companies in the same commercial position as other partners in the trade, and the burden of public service obligations on the railroad sector should not be transferred to other commercial operations without adequate compensation.

In the future, slurry pipelines should be a valuable alternative or complement to the railways. The critical issues are the ability to obtain a right-of-way for slurry pipelines, particularly in the United States, as well as adequate water availability in some regions. For international trade, slurry pipelines face additional constraints: the cost of storage at the loading terminals and the need to reduce the degree of moisture to lower the cost of seaborne transport. Failure to address these institutional and technical issues may further delay this promising transport possibility. Slurry pipelines might also have a beneficial effect of offering effective competition to railways where none now exists. Slurry pipelines should be allowed to compete fully with other modes of long-distance coal shipment, so that the market can enjoy the potential benefits from a technology that appears to offer an innocuous form of movement which is invisible, inaudible and efficient.

Infrastructure in Importing Countries

If the plans submitted by the IEA importing countries are realised, anticipated unloading capacity will be adequate to accommodate estimated imports. The access to the existing and planned ports would in most cases allow the use of large bulk carriers (100,000-120,000 dwt). A selected number of harbours would also be capable of handling larger vessels of 150,000 - 170,000 dwt, a size that can reasonably be anticipated in the late 1980s.

The degree of progress in implementing these plans varies substantially. Projects in Denmark, Germany, Ireland, the Netherlands and Spain are actively being developed; in Italy, Greece and Turkey projects are still in the planning stage. Major administrative difficulties relating to siting and the sharing of the financial burden exist in Italy. Greece and Turkey are very early in their planning and no precise timeframe has been revealed indicating when projects in these countries might mature. Portugal is considering the establishment of port facilities accessible to large coal carriers with a yearly capacity of 15 million tonnes. Three major sites are being examined in order to make a rational choice for the main reception port. Belgium already has substantial reception and transhipment facilities, but the access to ports is limited to coal carriers with a maximum of 85,000 dwt. As regards Japan, progress in developing reception capacities is under way, supported by the Energy Port Supporting Scheme.* Reception of coal will be handled by a combination of coal centres, public wharfs and private wharfs of the users. Currently, Japan is handling between 50 and 60 million tonnes of metallurgical coal and has substantial experience in reception of bulk commodities.

Continuing efforts are required from governments to maintain momentum in developing unloading facilities; temporary slowdowns in imports should not be allowed to affect long-term development programmes for harbour and reception facilities. Coal ports should be integrated in national harbour policies and possibly coordinated with neighbouring countries in order to avoid duplication of costly investment.

Oversizing the foundations of reception facilities as well as reserves of surface suitable for storage would appear to be a farsighted port development policy, because simple dredging of the access could then expand the rate of utilization of the existing installations, and allow the use of terminals by larger vessels.

The Coal Industry Advisory Board is currently conducting a comprehensive examination of infrastructure in both exporting and importing countries. Results are expected to be available in mid-1982.

E. Coal Prices and Economics

The substantial increase in oil prices in the period 1979-80 made coal strongly competitive in most IEA countries, particularly in the cement industry and electric power generation where the alternative fuel was oil. This situation changed the energy perspectives of many users and created high expectations among both consumers and producers. Conversion to coal and new coal-fired power plants were planned. As coal demand increased, some bottlenecks along the coal chain became apparent. A relatively difficult situation appeared in some exporting countries. This led to increases in coal prices and in transportation costs which, in turn, created some concern among consumers. Moreover the recent decline in oil prices has lessened the favourable differential between oil and coal and increased uncertainty on the part of potential consumers: to what extent are long term price trends for coal, relative to oil, likely to continue to be favourable?

Movements in coal prices over the last few years have reflected these special circumstances to a greater degree than longer-term trends. Recent increases in coal prices were in large part due to the rapidity with which overall demand grew and the resulting bottlenecks caused by the inability of existing

* The "Energy Port Supporting Scheme" is managed by the Ministry of Transport. The scheme provides a financial participation for building breakwaters, channels, anchorages and basins The port authority contributes an equal amount. The Ministry of International Trade and Industry also supplies soft loans for 40% of the cost of equipment of coal centres.

infrastructure to handle such a rapid increase. This general phenomenon was also reinforced by a number of particular events including the disruption of Polish supplies, a strike in Australia and anticipation of a strike in the United States. Recent price increases have nevertheless led to increased uncertainty on the part of many coal users.

Over the longer run, as the volume of trade in steam coal grows and infrastructure is further developed, coal prices should be largely determined by international market forces in which costs of production will play an important role. The high degree of competition in the United States domestic coal market should exert a restraining influence on coal prices in international trade, since US exports are not expected to amount to more than about 15% of domestic production.

It is difficult to say whether coal prices are likely to increase in real terms over the longer term. In many countries the general tendency will be for real costs of production to increase, but scope may well exist to offset this general tendency through greater rates of productivity increase or the achievement of substantial economies of scale in coal transportation, primarily through large vessels. It is equally if not more difficult to predict where real oil prices are likely to go, although all of the long-run scenarios developed by the IEA suggest that basic supply and demand trends point once again to upward pressure on oil prices in the late 1980s and early 1990s.

What is clear is that every effort must be made by those in the coal industry to minimize cost increases and strive for maximum productivity gains in order to protect and enhance coal's competitive position. Governments should also carefully assess the impact of fiscal measures on coal's ability to compete with other fuels.

The recent price experience shows that the international trading system for steam coal is still relatively fragile and has not yet developed to the point where fairly modest absolute changes in demand or supply can be easily absorbed without a major impact on price. More serious efforts need to be made in order to assure a smooth development of large basic trade in steam coal. This would help to avoid abrupt increases in coal prices through unanticipated short-term changes in supply or demand.

It is still true, even after the recent increases in coal prices, that steam coal prices (cif) are substantially below oil prices on a US$ per ton of oil equivalent basis. Annex 8 provides some general information on imported steam coal prices as of December 1981 for the two major importing areas: Europe and Japan. When compared with low sulphur heavy fuel oil spot prices, delivered coal prices show a substantial advantage: in Europe coal prices are equivalent to 40-55% of fuel oil price and in Japan 41-50%. This situation may not necessarily apply to all European countries and is in fact less attractive for the landlocked countries. Moreover, recent slight declines in oil prices have changed the picture to some extent. In addition, in discussing the economic advantage of coal over other fuels, it is not appropriate to make a comparison merely on a calorific basis; it must also be taken into consideration that coal use is accompanied by various indirect costs, such as costs of handling, investment in anti-pollution measures and procuring yards for stockpiling and ash disposal, and the relative advantage of coal over other fuels differs from industry to industry.

F. Environmental Considerations

Coal Production

While some regulations are felt to be less cost-effective than others, and variations are wide among operations at different sites, it is generally recognized that current reclamation standards and other environmental requirements do not seriously impede coal production or constitute a major portion of delivered coal costs. Reclamation is often costly, but the long-term social and economic benefits of permitting later use of mined lands as well as the short-term aesthetic benefits are also substantial, though more difficult to quantify. If these benefits are not achieved as a function of the mining operation, they may never be achieved, because they will probably demand a prohibitively large initial investment from those proposing later uses of the mined lands.

Coal Transportation

The environmental regulations pertaining to overland transportation are generally perceived as reasonable and have not caused substantial problems in areas where the coal industry has long existed. Problems may arise in new coal producing areas where coal was never carried in large quantities over long distances. Full appropriate information to the public before the actual start of operation is required, and the use of less intrusive transportation methods should be encouraged. For example, slurry pipelines, as well as providing an additional competitive source for inland coal transportation, could offer environmental benefits in populated areas that could be inconvenienced by heavily-used railroad crossings.

Coal Combustion

Coal combustion, and the related issue of waste disposal, are the subjects of most environmental concern in the coal fuel cycle. Public concerns about the expansion of coal use are focussed on the emissions of major air pollutants — sulphur and nitrogen oxides, and particulates — from the combustion process, and their effects on man and his environment, both short and long-range. This has caused a number of countries to adopt strict controls on sulphur content and amounts of gaseous emissions, or both, particularly from new installations. Although the situation varies depending on siting and regulatory factors, sulphur oxide emissions can generally be controlled wtih current technology and without the sacrifice of cost-effectiveness; the situation is less clear for nitrogen oxides. The result of this can often be that the development of new coal-fired plant, where it replaces existing oil-fired plant, leads to a significant net improvement in environmental quality — particularly with regard to sulphur emissions.

Other emissions, of more feeble intensity but even less well-understood as to their potential environmental and health effects, include trace metals and radioactive substances. The question of atmospheric build-up of carbon dioxide and its possible effect on global climate could also be relevant to the expansion of coal use, although these issues require a great deal of further study before policy conclusions can be confidently drawn.

Technologies exist for the control of most potential pollutants from coal combustion, and emerging technologies promise more effective control during the combustion process itself. The cost of pollution abatement is an important parameter currently under investigation by the OECD Environment Committee, and a Symposium on the economics of coal pollution abatement was hosted by the Netherlands Government under OECD patronage in May 1982*.

The relative prices of coal and oil are such that for new electricity generating plants, coal still remains competitive in general, even when a satisfactory degree of environmental protection is included. The situation is less clear for conversion or retrofitting of existing plants. Adequate environmental controls at reasonable cost are also less easy to ensure in the case of industrial boilers than in large generating plants. However, new technologies like atmospheric fluidised bed combustion show considerable promise for smaller boilers.

Problems which deserve further attention include:

- — whether government incentives, perhaps specifically connected to the costs of meeting environmental standards, might be desirable to encourage faster replacement of older, more polluting electricity generating plants in a climate of low electricity growth, capital cost inflation, and high interest rates;
- — harmonization of national and local standards to avoid conflicting or overlapping requirements which increase control costs but not benefits;
- — how information on appropriate technology for pollution controls during coal combustion, including emerging technologies, may be better disseminated to potential users, particularly in industry;
- — how to approach concerns of potential trans-boundary effects of coal-related pollution; and
- — environmentally satisfactory use or disposal of the wastes generated by coal combustion.

Regulatory Questions

There are two main issues causing concern in a number of countries:

- — first, the degree to which regulations should specify the means to attain specific standards or limits, or leave the user to select the most economic combination of technologies to comply with these standards or limits. This is particularly relevant to coal, where the most economic means of pollution control may be applied to various stages in the fuel cycle depending on the nature of the coal and of the system from mining through to waste disposal. It is generally preferable to restrict regulations, to the extent possible, to defining the end result required rather than the means. This allows maximum flexibility to develop appropriate approaches and new technology that can achieve derived objectives at the lowest feasible cost.

* The Proceedings of the Patten Symposium are expected to be published by the OECD during 1982.

— secondly, the integration and co-ordination of regulations. Pollution controls applied to the air medium produce solid or sludge wastes, which must then be disposed of in an environmentally sound manner. Regulations in most countries are developed and applied specifically and separately for air, water and land protection. Government departments are also organised along these lines. Failure to achieve appropriate coordination along the whole chain can lead to higher-than-necessary costs for compliance with environmental standards.

In general, achieving current environmental standards is not incompatible with increased use of coal. But, care will have to be taken by goverments, in cooperation with industry all along the coal chain, to ensure that regulatory frameworks allow those environmental standards to be met at minimum costs.

G. Summary of National Issues

A summary of the more important issues in individual IEA countries is presented below. Further details can be found, for each country, in Section III.

Australia: Appropriate solutions for labour supply and training are essential to the development programmes for coal mines. Timely development of railway infrastructure and ports are necessary but will depend upon firm commitments by foreign buyers. The coal export duty was extended in August, 1981, to include steam coal and this is giving rise to concern among current and potential buyers.

Austria: There is a need to diversify supply sources; the ability to do this depends greatly on the development of a port in the Adriatic Sea and an adequate inland transportation system from harbours in transit countries.

Belgium: Measures are necessary to assist conversion to coal in both electricity generation and industry. Access to ports should be deepened to allow for the use of large coal carriers.

Canada: As the financial requirements to develop new supplies and the associated infrastructure are quite large, firm commitments by foreign buyers are required to support timely mine and port development.

Denmark: Efforts should be intensified to support the penetration of coal wherever it can be used on a cost-competitive basis in accordance with the on-going overall heat planning.

Germany: Streamlining administrative and environmental procedures, regulating conversion to coal, especially in industry, would help increase coal use. The relaxation of import quotas, without damaging the current position of indigenous coal which should be maintained for improving the energy balance and for social and economic reasons, should be actively pursued where imported coal can be used to displace oil.

Ireland: Diversification of import supply is a necessity. Conversion to coal in industry should be more actively pursued. Interconnection of the electricity grid with the United Kingdom could increase security of electricity supply.

Italy: Implementation of the coal policy, which has been delayed for several years, is urgently required. Development of coal reception facilities is a prerequisite for construction of coal-fired power plants and expanded use of coal in industry. Urgent attention should be given to the inland distribution problems to supply the cement industry.

Japan: A continuing effort for developing reception capacities needs to be maintained to meet the requirements of the coal programme in a timely fashion. Accomodating environmental concerns to allow timely construction of coal-fired power stations should be aggressively pursued.

Netherlands: Political decision about the use of coal in electricity generation is needed. Appropriate solutions for meeting strict environmental standards have to be found.

New Zealand: Labour availability, and development of mines and infrastructure for inland transportation, represent bottlenecks to expanded use of coal.

Norway: Incentives would be required if the currently anticipated 15% share of coal in industry is to be lifted.

Portugal: Removal of subsidies for fuel oil would increase coal's competitiveness. Decisions to conclude long term contracts with foreign suppliers and to develop ports are required. The capability to finance the large-scale investments required is a major problem which requires urgent attention.

Spain: Development of coal-fired power plants with adequate reception facilities and commitment of users to long-term import contracts are areas where efforts must be pursued.

Sweden: The ambitious coal programme contained in the 1981 Energy Bill may be constrained by increasingly strict and changing environmental standards.

Switzerland: Greater use of coal is constrained by economic considerations stemming from, among other things, the landlocked geographical position of the country.

Turkey: Capital availability for investment in the whole coal chain is the major issue delaying further expansion of coal.

United Kingdom: Support to continued modernisation of coal production and adequate incentives to coal use are two areas where efforts should not be relaxed, especially in current economic circumstances which act as a constraint against the substitution of coal for oil.

United States: Critical issues affecting the pace at which domestic coal use will expand include the capacity plans and financial viability of electricity utilities; prices for competing fuels and particularly natural gas prices, which remain controlled; and environmental considerations, particularly as reflected in the Clean Air Act which is currently being reviewed. The coal movements currently envisaged for both domestic use and export will require the timely development of inland transportation capacity and ports. Particular attention should be given to slurry pipelines, dredging and ensuring the non-discriminatory treatment of coal destined for export by the railroads.

III NATIONAL COAL PROSPECTS AND POLICIES

A. IEA Member Countries

AUSTRALIA

Coal Use: Coal is currently the most competitive fuel for electricity generation, accounting for over 72% of primary fuel input into electricity generation. There is virtually no new non-coal-fired capacity under construction or planned. The State authorities are primarily responsible for electricity supply, and construction of oil-fired base load plants has been discontinued. In two states (Western Australia and Northern Territory), government programmes aim to convert existing oil-fired power plants owned by State electricity companies. In the industry sector, the government offers significant income tax concessions for eligible expenditures incurred on the replacement or conversion of eligible oil- and LPG fired plant to operate on alternative energy sources, such as coal. Considerable progress is being achieved in the non-ferrous metal industry and in the pulp and paper industry.

Energy Prices: Fuel oil prices reflect world oil prices and domestic crude oil is priced on a par with Saudi Arabian light crude oil adjusted for quality differentials and freight. Natural gas prices are not regulated by the Commonwealth Government; they are determined through long term contracts between private or state-owned utilities and producers and reflect the market conditions at the time contracts were negotiated. At present these prices are very low compared with the domestic oil price and with the gas price in Europe and in Japan. In recent negotiations the premium value of natural gas appears to be recognized but it is now substituting for fuel oil in many applications. At present coal, with prices calculated on an equivalent heat value basis, is the cheapest fuel available for industrial uses: it is between 35% and 40% of the natural gas price in the Melbourne and Sydney areas, respectively, and about 30% of the fuel oil price.

Coal Production: About 90% of mines are privately owned. Major growth is anticipated for the 1980s, specifically for export purposes. The major constraint on developing this sector is the current reluctance of consumers to enter into long term commitments with adequate pricing arrangements which would provide the basis for the investment to develop new mines and associated infrastructure. To obtain the projected production levels will require an enormous amount of capital, all of which cannot be raised in the domestic capital market. This scarcity of capital is being overcome through borrowing abroad and through the implementation of a foreign investment policy which welcomes foreign investment in Australia on specific terms and conditions. Policy on depletion of coal reserves is generally a matter for State authorities and most States reserve a portion of their reserves for electricity and/or industrial use. The Commonwealth Government has provided major income tax concessions to encourage mineral exploration and development, especially in the 1976-77 budget. This very generous system of write-offs was reduced somewhat in the 1981-82 budget, in view of the fact that the mineral sector has now entered a period of strong growth which is expected to continue for some time.

The Commonwealth Government welcomes capital investment in production and transportation facilities from overseas, including from potential buyers, subject to its foreign investment policy. The objective of this policy is to have, as a general rule, at least 50% Australian equity and control in major coal projects. Reclamation of mined lands is carried out in accordance with the provisions of each State's Mining and Environment Acts.

Availability of engineers and skilled manpower is a major constraint that could be overcome by a long-term recruitment and training programme. The Commonwealth and State/Territory governments have recently agreed on comprehensive policies of close consultation with industry, early identification of labour and skill needs, recruitment sources, substantial training effort by contractors and developers and supplementary assistance from government manpower programmes.

State governments are responsible for imposing royalties on the production of coal within their borders. The royalty rates vary from state to state with the coal exporting states imposing higher rates (*New South Wales:* A$ 1.70 per ton; *Queensland:* open cut not to be consumed inside the state, 5% of free on rail (FOR) value; underground not to be consumed in Queensland, 4% FOR value). All coal produced is levied with an excise duty at a rate of A$ 0.15, 10 cents of which is paid into the coal mining long-service leave fund and 5 cents to support RD&D. The coal export duty imposed by the Commonwealth Government was extended in August 1981 to include steaming coal, at a rate of A$ 1.00 per ton.

Transportation: The increase in coal export demand will require additions to infrastructure capacity (rail and port facilities). The largest immediate problem is to obtain firm commitments from consumers. The government is seeking such commitments and has also initiated consultations with the States to ensure the timely development of infrastructure. New port facilities have been commissioned in Gladstone and work is in progress on development of Dalrymple Bay and expansion of facilities at Port Kembla, Newcastle and Balmain. Harbour deepening at Gladstone is underway. Currently the most efficient land transport is by rail but the remoteness of mines may cause the provision of facilities to be extremely costly and difficult to justify without long-term throughput.

Research into coal slurry lines is proceeding with one pilot plant at Newcastle. If this system is introduced on a commercial scale, rights of way, water use and tariff policies could become issues. A difficult matter requiring constant scrutiny is the issue of tariffs. In certain instances (Queensland) tariff regulations of the State-owned railway penalize the transport of coal for export to provide subsidies to other railway transport such as public transportation and this is detrimental to coal's competitiveness.

Trade: The government seeks to ensure that, in the national interest, exports of coal take place under terms and conditions which are in line with those in the world market. To this end, approval is required for the export of coal but the government tries to minimize interference in normal commercial processes, leaving negotiations to the commercial interest involved. Export control is regarded as a reserve power against the event of the national interest being threatened. This power is not used to interfere with long-term contracts for international trade in coal once they have come into force.

Environment: State authorities are largely responsible for environmental issues associated with coal use. Environmental policies are co-ordinated through the Australian Environment Council, comprising Commonwealth and State Ministers with responsibilities for environmental matters.

Mining and transportation activities carry environmental risks for which adequate control measures have been designed: a practical example is the withdrawal of the planned development of an open-cut mine adjacent to a partly constructed power plant following an adverse environmental impact report. Coal will now be railed and trucked to the power station from mines further afield. The combustion of domestic coal is not environmentally damaging because of its low sulphur content.

Research is being carried out into waste disposal techniques, such as the burning of coal with high ash content.

RD&D: RD&D activity is growing and includes investigations covering all aspects of coal production and utilization. Funding under the National Energy Research, Development and Demonstration Program (NERDDP) is provided from the Coal Research Trust Account based on a 5 cents per tonne levy on coal production and from appropriations to the Department of National Development and Energy. Commonwealth Government funding of energy RD&D under NERDDP is US$ 15.5 million in the 1980-81 budget. (US$ 10.0 million in 1979-80). This funding is additional to State Government and industry-financed coal research.

Coal mining research is mainly directed at developing underground mining systems and investigating environments that will allow higher seam recovery as the inevitable trend towards deeper mines evolves. Aspects of surface mining, beneficiation and transportation are also under study.

Coal utilization research is largely concerned with combustion and conversion aspects of Australian coals. In the combustion area, investigations include aspects of the behaviour of bituminous and low rank coals in pulverized fuel firing, fly ash precipitation and collection, and fluid bed combustion. Research in coal conversion is primarily concerned with studies on the liquefaction potential of Australian coals and is intended to facilitate selection of appropriate technology for commercial application by the private sector.

Summary of Specific Issues

Coal production, which is projected to increase about two-and-a-half-fold from 1980 to 1990 is strongly encouraged by Commonwealth and State governments.

A major constraint on the production side is a short supply of manpower: both the Commonwealth and State/Territory governments have agreed on comprehensive labour policies in close consultation with industry. Particular attention regarding foreign workers and appropriate training programmes will be required. An improvement in the industrial relations in both coal production and infrastructure areas would also contribute to help expansion of the long-term commitments for coal exports. A substantial increase in coal exports is foreseen: from 30 Mtoe in 1980 to 78 Mtoe in 1990. This projected development will require firm commitments by foreign buyers (mainly from Europe and Japan) in order to allow producers to make the necessary investment in mining capacity and infrastructure. The introduction in 1981 of an export duty on steam coal (A$ 1.00/tonne) is giving rise to concern.

AUSTRALIA

Key Solid Fuel and Energy Indicators
(Mtoe)

	1973	1979	1980	1990	Average Annual Growth Rates (%) 1973-1980	1980-1990
Production						
Total Energy Production	69.6	93.3	92.8	190.6	4.2	7.5
Solid Fuel Production	43.2	57.8	59.0	131.8	4.5	8.4
Solid Fuels as % of Total	62.2	61.9	63.5	69.0		
Trade						
Coal Exports	18.2	26.3	27.9	78.1	6.2	10.8
Coal Imports	0	0	0	0	n.a.	n.a.
Requirements						
Total Primary Energy Requirements	58.2	76.8	75.8	108.5	3.8	3.6
Solid Fuel Requirement	25.0	31.6	32.7	53.4	3.9	5.0
Solid Fuels as % of Total	42.9	41.2	43.1	49.2		
Sectoral Consumption of Solid Fuels and Comparison to Oil						
Solid Fuel Use						
Industry [1]	7.4	7.6	7.2	13.1	0.1	6.1
Iron and Steel	3.1	3.4	3.0	n.a.	- 0.7	n.a.
Other	4.0	4.1	4.1	n.a.	0.1	n.a.
Residential and Commercial [2]	0.7	0.8	0.7	0.7	1.5	- 0.1
Electric Generation [3]	13.9	20.8	22.4	36.7	7.1	5.1
Solid Fuel Share (%)						
Industry [1]	44.2	37.8	35.2	37.4		
Iron and Steel	73.6	57.9	60.8	n.a.		
Other	34.0	31.4	29.5	n.a.		
Residential and Commercial [2]	9.5	7.9	7.8	5.4		
Electric Generation [3]	75.1	76.3	78.5	80.1		
Oil Share (%)						
Industry [1]	30.2	27.0	25.5	12.9		
Iron and Steel	25.8	28.7	9.9	n.a.		
Other	31.4	31.8	29.3	n.a.		
Residential and Commercial [2]	54.2	38.4	40.5	32.3		
Electric Generation [3]	2.3	2.8	3.9	2.2		

1. Excludes non-energy use.
2. Includes Public and Agricultural Use
3. Represents fuel input to electric generation

AUSTRIA

Coal Use: In November 1981, the electric utilities decided on their coordinated construction programme for 1981-1990 which shows that almost all of the large thermal power plants to be constructed to 1990 will be coal-fired and the other main power source will be hydro. Investigations of environmental objections may slow down the process, as is currently the case with the construction of two thermal power plants to be fired with coal imported from Poland. The decision to proceed with these plants was taken in mid-1981. The first is expected to start production in 1985-86 and the second in 1986-87. In these planned power plants about 3.5 Mtce will be burned yearly and in addition 0.5 Mtce will be needed to operate existing plants. The share of solid fuels for thermal generation, which amounted to 28% in 1980, is estimated in latest projections to be 50% by 1990. The delays in the construction of new thermal power plants and hydro power plants together with the ongoing consequences of the mothballing of the completed nuclear plant, which has a potential production of about 10% of current electricity needs, will require additional electricity imports. Present construction work to allow the exchange of electricity with other countries — both UCPTE area and Eastern Europe — may need to be expanded.

Electricity generation is closely tied to DH and CHP undertakings. Such projects are allowed to constitute tax free reserves up to 50% of annual profits to be used, for example, for expenditures to convert existing oil-fired plants into multi-fired power stations wherein solid fuel must be included.

In the industry sector, oil displacement by coal is low and essentially governed by the economics: although coal has a definite cost advantage, higher capital and operation costs of coal-fired equipment as well as the prohibitive cost of conversion of oil-fired boilers act as a major constraint on possible conversion of boilers constructed to burn oil. Coal, to compete, needs to be priced at least 20% to 30% below oil prices (on an equivalent heating value basis) but the upper limit of this range is hard to reach due to high transport costs. Oil and gas substitution cannot be enforced by law. Currently, solid fuels are expected to reverse their recent decline and grow to 2.6 Mtce in 1985, remaining at this level until 1990. Further investigation into measures and incentives to promote switching to coal in industry appear necessary.

Energy Prices: The prices of major energy products are subject to price control through a series of complex and non-uniform systems administered by the various levels of government and by the Price Commission. These are aimed at avoiding shocks to the economy through sudden large price variations. Currently fuel oil prices are relatively low and coal prices, which are exempted from control, are not very competitive.

Coal Production: Domestic production includes only brown coal at relatively modest levels (1.2 Mtce in 1980). During 1980, AS 22 million* (US$ 1.4 million) were invested for prospecting, exploration and development, of which 40% was funded by the federal state and the Länder. No decision has yet been made about the planned new venture with Hungary for the exploitation of a lignite deposit straddling the border, pending assessments of exploration results in another area of the country (Carinthia).

Transportation: Because of its landlocked position, imported coal into Austria carries high transportation costs. For current and anticipated imports from Poland, an expansion of the carrying capacity of the railway between Poland and Austria via Czechoslovakia has been completed. Inland transport of coal imported from other countries will, at least until 1985, be managed by railway or via the Danube from the Black Sea, as well as from the West to the extent possible since the finalization of the Rhine-Main-Donau canal is uncertain at this time. An expansion of port capacity in the Northern Adriatic region is another possibility for handling coal imports from overseas. This would require upgrading the railway connection in neighbouring countries between the ports and Austrian consuming areas.

Trade: Because of limited domestic production, Austria has to expand its coal imports. In June 1980 an agreement was reached with Poland for the delivery of 20.4 million tons of steam coal with an option for additional 10 million tons over a period extending to the year 2004.

Diversification of future coal imports is a priority objective for both government and coal users. Austrian firms are negotiating with the United States, Australia, and other countries, in order to obtain shares in coal producing undertakings abroad to secure coal supplies.

Environment: After an analysis by the Austrian Academy of Sciences of the emission of dust and sulphur dioxide by the two coal-fired power stations in lower Austria, it was decided to equip these plants with desulphurization equipment for flue gases.

It has also proved necessary to install the same kind of equipment for all planned lignite-fired stations in the future (e.g. Voitsberg 3 near Oberdorf).

RD&D: Research to improve efficiency of coal combustion technology is done at the Technical Universities of Vienna and Leoben (Styria) and, on a limited scale in steel firms active in boiler construction.

Summary of Specific Issues

Austria is currently too heavily dependent on Poland and needs to diversify its supply sources for imported coal. A major goal is to improve inland transportation from ports to consuming areas in order to alleviate the burden of its landlocked position and maintain an adequate economic margin in favour of coal at the point of consumption. This will require bilateral agreement with neighbouring countries (for example, with Italy and Yugoslavia for the development of ports in the Northern Adriatic region). Waste disposal (ashes and sludge) could raise difficulties, particularly in densely populated areas.

* AS 1 = US$ 0.63

AUSTRIA

Key Solid Fuel and Energy Indicators
(Mtoe)

	1973	1979	1980	1990	Average Annual Growth Rates (%) 1973-1980	1980-1990
Production						
Total Energy Production	10.4	11.6	11.3	12.1	1.2	0.7
Solid Fuel Production	1.3	1.5	1.5	1.0	1.9	- 4.0
Solid Fuels as % of Total	12.6	12.5	13.2	8.2		
Trade						
Coal Exports	0.1	0	0	0	- 13.6	n.a.
Coal Imports	3.0	3.0	2.9	4.1	- 0.4	3.4
Requirements						
Total Primary Energy Requirements	24.3	27.1	26.8	35.0	1.4	2.7
Solid Fuel Requirement	4.3	4.4	4.4	5.1	0.4	1.4
Solid Fuels as % of Total	17.5	18.4	16.4	14.4		
Sectoral Consumption of Solid Fuels and Comparison to Oil						
Solid Fuel Use						
Industry [1]	1.4	1.7	1.6	1.8	1.1	1.2
Iron and Steel	1.2	1.5	1.4	n.a.	2.2	n.a.
Other	0.2	0.1	0.1	n.a.	- 2.3	n.a.
Residential and Commercial [2]	1.5	1.6	1.6	1.1	0.8	- 3.4
Electric Generation [3]	0.7	0.8	0.8	1.8	1.3	9.1
Solid Fuel Share (%)						
Industry [1]	18.4	24.0	22.9	20.0		
Iron and Steel	51.8	60.5	59.0	n.a.		
Other	3.8	3.7	4.2	n.a.		
Residential and Commercial [2]	24.6	22.0	23.7	10.7		
Electric Generation [3]	10.2	8.1	8.0	14.4		
Oil Share (%)						
Industry [1]	45.9	31.0	31.9	25.8		
Iron and Steel	25.0	17.5	18.7	n.a.		
Other	56.2	46.2	45.9	n.a.		
Residential and Commercial [2]	49.7	47.4	41.4	51.1		
Electric Generation [3]	12.1	14.5	15.5	6.4		

1. Excludes non-energy use.
2. Includes Public and Agricultural Use
3. Represents fuel input to electric generation

BELGIUM

Coal Use: In conventional electric power plants, the share of coal in total fuel inputs increased from 28% in 1973 to 36% in 1980. A supplementary programme for the gradual conversion of oil-fired generating capacity (750 MW) was enacted in October 1979 and will result in a reduction in oil consumption of about 1 million tons/year by 1983. No new coal-fired power station is currently planned, but no decision has yet been taken on a proposed new plant designed to burn domestic steam coal (Campine area). The electric utility companies agree to use all steam coal produced by the mining industry at a price comparable to the average price of all other fuels used (imported coal and oil) through a system called "pool of calories" centralising all the data referring to fuel input in the electricity industry.

In the industry sector, switching from fuel oil to coal is planned to stabilize oil consumption at current levels throughout the 1980s. Technically there are no obstacles to total conversion but the drawbacks are economic. Total costs (financial, operating, maintenance, etc.) involved in switching to coal are higher than those associated with the use of fuel oil or gas in existing installations. Increased financial incentives could speed up this conversion process, which presently depends on persuasion through bilateral talks. Certain incentives are offered to the agriculture and residential sectors for converting boilers to coal or gas in the first instance and to stimulate the purchase of new automatic coal-burning boilers in the latter. District heating through co-generation is used only on a small scale although feasibility studies indicate a substantial potential (3530 MW) and an initial phase (100 MW) with five sites has been started. Recent studies indicate that the extension of the coal-based CHP capacity in industry could supply about 80% of industrial heat requirements by 1990.

Energy Prices: Energy prices and more specifically heavy fuel oil are determined by regulation mechanisms which take into account increases in world market prices. The rate of VAT on heavy fuel oil is 17% while for coal the rate is 6%.

Coal Production: The present level of production (6.3 Mtce in 1980) is expected to remain almost unchanged to 1990 and the subsidy system now employed to offset higher costs is planned to be maintained. The subsidies are supplied from the budget and granted basically to the mining industry helping to bridge the gap between production costs and prices obtained on the market. The amounts are substantial (US $47 per tonne produced in 1980) but it is considered unavoidable by the government for both social and regional reasons, as well as contributing to security of supply. Presently mined concessions are located in Limbourg. The concession holder has carried out exploratory drilling during the last two years and the data obtained will be studied to help determine the extent to which new coal reserves might be exploited in the future.

Transportation: The present port handling capacity and inland transportation system are adequate but additional handling facilities may be required if coal use is to expand substantially. A constraint which has not yet been overcome is the need for dredging of the access to Antwerp, which is currently limited to coal carriers of 85,000 dwt although the locks and unloading facilities could accomodate larger vessels.

Trade: There are no trade barriers restricting coal imports. An increasingly large volume of imports will be necessary to satisfy growing requirements for steam coal. Additional new supply contracts should be negotiated to start actual delivery by the end of the current decade especially if no positive decision on nuclear expansion is reached soon.

Environment: Use of coal in power plants has to meet mandatory requirements regarding maximum emission of dust in stack gas. Efficiency tests are foreseen for dust extractors. Other solid fuel users in the industry and residential sector have to comply with smoke abatement requirements within special protection areas and maximum contents of dust in stack gas. The recovery, for burning in electric power plants, of solid fuels from coaltipples is made under strict environmental control regarding dust, noise and land reclamation.

RD&D: The government has allocated resources to a Belgian-German underground gasification experiment which started in 1979 and is scheduled to continue until 1986. This experiment aims at producing gas suitable for fueling a gas turbine power station. The government is allocating funds for a coal hydrogenopyrolysis system and the EEC is granting support for fluidized bed techniques in the industry sector.

Summary of Specific Issues

Timely completion of the conversion of oil-fired generating capacity to coal as well as an early decision on the proposed new plant based on indigenous coal (Limbourg) are required.

Incentives for the conversion from oil to coal in the industrial sector are needed to help overcome the cost of non-productive capital investments. Government should find more appropriate measures to assist the conversion process, and give greater support to DH/CHP plants.

Timely dredging of the access to the coal unloading installation in Antwerp is needed to yield the full benefit of lower transportation costs by use of larger vessels.

BELGIUM

Key Solid Fuel and Energy Indicators
(Mtoe)

	1973	1979	1980	1990	Average Annual Growth Rates (%) 1973-1980	1980-1990
Production						
Total Energy Production	6.4	7.1	7.7	12.8	2.7	5.2
Solid Fuel Production	6.2	4.3	4.4	4.2	- 4.6	- 0.5
Solid Fuels as % of Total	96.6	60.8	57.6	32.8		
Trade						
Coal Exports	0.6	0.8	0.9	0.5	5.6	- 5.7
Coal Imports	5.9	7.8	8.0	11.3	4.5	3.4
Requirements						
Total Primary Energy Requirements	46.7	48.6	46.8	56.7	0.0	1.9
Solid Fuel Requirement	11.7	11.3	11.6	15.0	- 0.2	2.6
Solid Fuels as % of Total	25.1	23.3	24.8	26.5		

Sectoral Consumption of Solid Fuels and Comparison to Oil

	1973	1979	1980	1990	1973-1980	1980-1990
Solid Fuel Use						
Industry [1]	5.7	5.0	4.9	6.6	- 2.3	3.1
Iron and Steel	5.0	4.0	3.8	n.a.	- 3.8	n.a.
Other	0.7	0.9	1.0	n.a.	6.1	n.a.
Residential and Commercial [2]	2.4	1.3	1.1	0.5	- 10.6	- 7.4
Electric Generation [2]	2.8	4.0	4.6	6.8	7.2	3.7
Solid Fuel Share (%)						
Industry [1]	31.8	31.0	33.3	35.5		
Iron and Steel	69.9	69.4	71.7	n.a.		
Other	10.5	15.3	19.1	n.a.		
Residential and Commercial [2]	18.0	8.8	8.0	3.3		
Electric Generation [3]	28.4	33.4	35.3	37.1		
Oil Share (%)						
Industry [1]	36.8	28.4	24.7	25.8		
Iron and Steel	10.9	5.8	3.6	n.a.		
Other	51.9	37.4	32.3	n.a.		
Residential and Commercial [2]	63.6	57.1	52.9	47.4		
Electric Generation [3]	50.2	31.2	30.0	10.1		

1. Excludes non-energy use.
2. Includes Public and Agricultural Use
3. Represents fuel input to electric generation

CANADA

Coal Use: The electric utilities, which are mainly based on hydro, coal and nuclear, anticipate an increase in the share of coal used for electricity generation, primarily near opencast mines in order to avoid long distance railway transport. The National Energy Programme (NEP) provides grants to finance 75% of the cost of conversion to coal for electrical utilities in the Atlantic Provinces, provided the design is environmentally acceptable. A detailed study of the possible means and costs of converting Canada's largest (1005 MW) base load oil-fired power station (Coleson Cove in New Brunswick) to coal firing is under way and should be completed by early 1982; this power plant represents 12% of the total oil-fired installed capacity.

Oil-fired electricity generating capacity represents 10% of total installed capacity and this share is expected to come down to about 7.5% in 1990. Coal-fired generating capacity is expected to increase by 29% over the decade. Construction and planning of coal-fired capacity is under way in Alberta, Saskatchewan, Ontario and Nova Scotia. British Columbia has completed the evaluation of the Hat Creek coal deposits for an eventual generating capacity of 2000 MW and has begun a feasibility study of combining electricity generation with conversion of coal to liquid or gaseous fuels.

Coal's competitiveness for electricity generation varies from province to province. In Alberta and Saskatchewan, coal is the economically preferable fuel for base, intermediate and even peak load generation. In Ontario coal is competitive with nuclear and hydroelectric power for intermediate and peak loads, but not for base load. In Manitoba, coal is competitive with hydraulic sources for emergency and peak loads only. In New Brunswick and Nova Scotia, coal's major competitor has been oil: coal has recently become competitive with residual oil and is expected to become more so as oil subsidies are eased. Federal loans will continue to be available to finance interprovincial electrical interconnections; this could increase the capacity to provide hydroelectric power from Quebec and Labrador to the Maritime Provinces.

In the industrial sector the major competitors of coal are heavy fuel oil and gas, with gas having an economic advantage in most regions because of controlled prices and transportation differentials. The goal established by the federal government in the National Energy Programme is to reduce by 1990 the use of oil to 10% of energy requirements in the residential/commercial and industrial sectors. Several programmes are now under development and will encourage users to substitute more plentiful sources of energy for oil with the choice left to the consumer between a wide range of alternatives to oil including natural gas, propane, electricity, wood, solar and coal. Conversions to some of these alternatives are not eligible for grants in some provinces at provincial request. Substitution of coal for oil will be achieved through direct replacement wherever economically and technically feasible, through coal-liquid mixture combustion, through use of coal for fuel-processing energy that can save oil, through coal-based oil synthesis and through produced chemicals such as methanol that might be used directly or indirectly as transformation fuel. The use of coal and nuclear reactors as a source of energy to upgrade oil sands and heavy oils is envisaged. Studies to investigate the economical and technical feasibility of this use are under way.

Energy Prices: Under the federal system, jurisdiction over energy pricing is divided between the provincial and the federal governments. The provinces have jurisdiction by virtue of their ownership of resources, while the federal government has the power to regulate interprovincial and international trade. It is therefore considered preferable to have energy pricing determined by negotiation between the two levels of government. The federal government, however, has the ability to set prices when the energy is sold for use outside the producing province. Agreements between the Government of Canada and the provinces of Alberta (87% and 90% of Canadian oil and gas production), British Columbia and Saskatchewan were reached in September and October 1981. The agreements provide for a resolution of the conflict over sharing of oil and gas revenues and affirm the commitment to NEP measures designed to encourage the development of new oil and gas supplies. The price for oil discovered prior to 1 January 1981 will increase at six-monthly intervals to reach in July 1986 a level of $C 57.75 per barrel subject to the condition that this is no greater than 75% of the average cost of imported oil delivered at Montreal. Oil discovered subsequently, oil recovered by enhanced recovery schemes and production from oil sands plants is foreseen to reach a price level of $C 77.48 per barrel, with the actual price to be based on a two-year moving average floor price which would not exceed 100% of the world price. These arrangements continue the long-term aim of the NEP to keep the blended price of domestic and imported oil below the world price, particularly during periods of sharp or rapid world oil price escalation. Consumer prices for oil products will nevertheless increase quite rapidly because of higher wellhead prices for old oil, increased taxes on oil consumption, and the increasing share of "new oil" in Canadian supplies over the decade and beyond. For natural gas the agreement with the Western provinces provides that the commodity will be priced at approximately two-thirds the price allowed for oil, calculated on conversion to a Btu equivalent.

This pricing regime will act as a substantial incentive for consumers to convert from oil to natural gas and contribute to reducing oil consumption in the residential and industrial sectors to no more than 10% of total energy needs. The federal government will continue to levy a Petroleum Compensation Charge on all oil consumption to finance the differential cost of importing oil, as long as necessary.

Coal Production: Coal production is generally vested in the private sector in western Canada and largely under control of crown corporations in the East. Private coal producers receive the same tax treatment as the mining industry in general. Royalties are virtually minimal in British Columbia (3.5% on fob value) and Nova Scotia (C$ 0.28 per metric ton), but they are higher in Saskatchewan (15% on selling price) and are related to profitability in Alberta (5% of the market price determined by the government taking into account rate of return). No federal royalties are imposed.

Provincial governments control the disposal of coal and the licensing of mines. Coal resources are adequate to supply an increasing domestic demand and an expanded export market for both steam and coking coal. A major difficulty is that the main coal resources are located in the middle west of the country, separated from the potential markets in the east by long distances and from the exporting terminals by difficult terrain. Transportation is thus a major issue.

The Foreign Investment Review Act requires that the Foreign Investment Review Agency examine any new foreign investment in Canadian ventures, including coal mines and transportation facilities, to determine whether such investment is of ultimate benefit to the country. The Foreign Investment Review Agency has recommended authorization of most coal-related proposals.

Foreign investment in coal production facilities is below 40% and maintenance of this share is in line with Canadian energy policy. The existing policy is one of minimum intervention leaving the coal industry to the play of commercial forces with demand dictating coal development.

Transportation: Forecasts indicate that increasing demands for bulk commodities for both domestic demand and export will require major expansion of railway capacity in western Canada in the 1980s. Both of Canada's major railway companies have initiated investment programmes to upgrade their facilities, equipment and systems. Expansions of port facilities are under way at Sydney (Nova Scotia) and at Roberts Bank (British Columbia), where handling capacity is scheduled to double to 22 million tons/year by mid-1982. A new terminal at Ridley Island (British Columbia) is being developed with a contemplated capacity of 12 million tons/year. These installations will ensure adequate capacity to handle projected increases in coal movements in the coming decades. A major issue lies in the economies of scale for initial movement on new rail lines, on upgraded rail lines and through new coal terminals. The primary transportation constraint in developing new coal production regions is the difficulty in obtaining adequate throughput in the early years to cover the capital costs of new rail lines and ports. No major alternative transportation modes presently exist for the inland movement of coal. The movement of natural resources in Canada has traditionally contributed to the expansion and upgrading of national transportation systems. Future expansion of the coal industry in Western Canada, especially in northeastern British Columbia, will be tied to upgrading and expansion of the national and regional transportation systems. These changes will benefit not only the coal industry, but also existing and potential users of the transportation system.
One issue is, however, the distortions in rate setting such as the very advantageous mandatory rates for rail transport of grain which can lead to higher rates for other commodities.

Trade: The concept of increasing coal trade to meet growing world energy requirements is actively supported and the federal government does not set limits on exports in either steam or coking coal.

The provinces, in general, give first preference to intra-provincial use, second preference to use in other provinces and third preference to export but the size and quality of the coal deposits indicate that there is no concern, at present, over domestic requirements constraining foreseeable export potential.

Environment: The federal government is responsible for ensuring that environmental impacts are evaluated and determined to be of an acceptable nature in the case of projects receiving federal support. In the general case, responsibility for environmental matters falls within the domain of the provinces. Past coal production practices had substantial environmental impact and led to public concern that mining areas be rehabilitated as fully as practicable when activities cease. This rehabilitation is primarily a provincial responsibility and the current reclamation practice seems to be satisfactory.

The problem of acid precipitation in eastern North America, resulting in large part from the use of steam coal in power plants, has led to detailed studies on advanced combustion technologies and other means to contain sulphur dioxide and nitrogen oxide.

RD&D: A major issue for future coal development is the development of clean-burning systems for using coal of variable quality. The main thrust in the development of such systems is a series of pioneer commercial fluidized bed combustion units of successively larger size leading ultimately to pressurized fluidized bed combustion at electric utility scale. Construction has started (Prince Edward Island) on the first commercial fluidized bed boiler to use high sulphur coal and contain atmospheric emissions; the target date for completion is the 1982-83 heating season. Projects for industrial process steam and coal drying, using fluidized bed combustion to burn coal washery rejects are in their early stages. Proposals for coal liquefaction are at the stage of matching the country's coal resources with known conversion processes in order to select the best processes suitable for the different regions. Currently there are four projects, each concerning a specific coal, at the feasibility study stage. The projects, which include assessment of product yields by liquefying coal in small continuous units, are expected to reach conclusion as to economic and technical feasibility by 1983.

Programmes to investigate the economical and technical feasibility of the use of coal as a source of energy to upgrade oil sands and heavy oils are underway.

Summary of Specific Issues

While the Canadian coal industry has expanded rapidly in the last few years, coal policy concerns have not received the same high priority given the larger oil and gas problems. On the demand side, the existing policy is to leave the coal industry to operate commercially. Increasing conventional domestic use will expand rapidly in Western Canada especially at new minemouth power plants. The resource base for domestic and export markets is large but expansion of productive capacity generally depends on firm long-term commitments by end users. The expansion of coal terminal infrastructure is well underway at the ports of Vancouver and Prince Rupert and recent initiatives by government and industry should ensure adequate long-term rail system capacity to handle increasing coal exports in the 1980s and 1990s.

CANADA

Key Solid Fuel and Energy Indicators
(Mtoe)

	1973	1979	1980	1990	Average Annual Growth Rates (%) 1973-1980	1980-1990
Production						
Total Energy Production	223.0	242.8	242.3	283.6	1.2	1.6
Solid Fuel Production	11.5	23.2	25.6	41.9	12.0	5.1
Solid Fuels as % of Total	5.2	9.6	10.6	14.8		
Trade						
Coal Exports	7.6	8.8	9.9	15.7	4.0	4.7
Coal Imports	10.4	11.4	10.3	10.3	0	0
Requirements						
Total Primary Energy Requirements	190.6	222.9	228.2	269.1	2.6	1.7
Solid Fuel Requirements	15.1	24.3	26.2	36.5	8.2	3.4
Solid Fuels as % of Total	7.9	10.9	11.5	13.6		
Sectoral Consumption of Solid Fuels and Comparison to Oil						
Solid Fuel Use						
Industry [1]	5.7	10.8	11.1	17.1	10.0	4.4
Iron and Steel	3.7	4.2	4.1	n.a.	1.5	n.a.
Other	2.0	6.6	7.0	n.a.	19.3	n.a.
Residential and Commercial [2]	0.4	0.1	0.1	0.1	- 14.8	- 2.5
Electric Generation [3]	8.2	13.0	14.7	19.3	8.7	2.8
Solid Fuel Share (%)						
Industry [1]	14.2	20.4	21.1	21.9		
Iron and Steel	62.5	57.8	57.7	n.a.		
Other	6.9	19.0	20.7	n.a.		
Residential and Commercial [2]	0.9	0.3	0.3	0.2		
Electric Generation [3]	11.9	15.5	16.5	16.3		
Oil Share (%)						
Industry [1]	35.6	30.5	27.2	16.7		
Iron and Steel	9.0	12.3	11.0	n.a.		
Other	43.0	27.6	25.1	n.a.		
Residential and Commercial [2]	48.5	38.9	37.0	16.4		
Electric Generation [3]	4.1	3.9	3.4	2.1		

1. Excludes non-energy use.
2. Includes Public and Agricultural Use
3. Represents fuel input to electric generation

DENMARK

Coal Use: Coal accounted for 22% and 32% of primary energy consumption in 1979 and 1980 respectively. Coal is mainly used in power stations for generating electricity and heat. Of total coal consumption, the power stations accounted for 86% and 90% in 1979 and 1980 respectively. The remainder is used in processing industries (a little less than 10% of total coal consumption in 1980), district heating plants and in a gas-work. Almost all coal consumption consists of steam coal.

In electricity generation the switch from oil to coal has been remarkable: the share of coal increased from 21% in 1972 to 62% in 1979 and 83% in 1980; by 1990 the coal share in electricity generation is expected to be 90%. No specific measures have been necessary to enforce the switch from oil to coal because the increasing gap between coal and fuel oil prices has been a sufficient incentive. All new power stations will be built for coal-firing, with back-up capabilities for oil-firing.

In the Spring of 1980 a new company, Danish Power Consult Ltd., was created. The aim of the company, for which the power stations work as consultants, is to promote Danish know-how concerning switching from oil to coal and concerning the use of coal of different qualities.

When planning new power generating capacity, the advantages and disadvantages of large coal-fired units as opposed to smaller decentralized units are discussed. This discussion should be seen in the context of the implementation of the Danish Heat Plan, where the use of natural gas and district heating also play important roles.

Coal in the industry sector, including tile works etc., accounted for 17% of total energy use in 1980 in that sector. The main consumption takes place in the cement industry and in the paper and sugar factories. The increase in the coal consumption of the industry sector is uncertain; present estimates indicate that by year 2000 coal may account for 22% of total industrial energy use.

Energy Prices: Prices, including fuel oil, are generally determined by market forces. Prices for electricity are determined by costs (including replacement costs). According to the Heat Supply Act, future heat and gas prices will likewise be determined by actual costs; a Gas and Heat Pricing Committee has been established.

Coal Production: No indigenous coal deposits.

Transportation: All coal is imported by ship. Most of the coal for power generation and industry is landed in ports owned by the utilities, which are usually located close to the ports.

The Ministry of Public Works is preparing a comprehensive plan for the Danish port system, which will take into consideration the foreseen increase in Danish coal imports. The possibility of creating a coal centre which could serve Scandinavian countries is under consideration.

At present the largest coal harbours owned by the power stations can accommodate fully laden ships up to 100,000-110,000 dwt. There are two such harbours, one in Jutland and one on Zealand. It has recently been decided by the boards of the power stations to enlarge the two harbours so that fully laden ships up to 170,000 dwt can be received. These projects are expected to be completed by 1984. In the port of Aarhus a terminal is under construction for receipt of fully laden ships up to 60,000 dwt and partially loaded ships of up to 120,000 dwt.

Trade: Coal imports are unrestricted and are typically undertaken by the end-users themselves. Only about 1% of total imports is undertaken by commercial importers for re-selling to, among others, tile-works, households, farmers, market-gardens and district heating plants.

Total Danish imports rose from 2.3 Mtce in 1972 to 9 Mtce in 1980 and are expected to reach 11.6 Mtce by 1990. The increase in total imports has been characterized by a rapid increase in overseas imports. While imports in 1973 and previous years came from European countries only, overseas imports in 1979 accounted for 43% and in 1980 for 52% of total imports.

Based on a decision of Parliament, the government has recommended that power stations diversify imports geographically. In March 1980, a Committee of Coal Supplies was set up with representatives from the power stations, the Ministry of Energy and other involved ministries. There is a consensus in the Committee that one should aim at long-term contracts with the big coal-producing/exporting countries which, like Denmark, are members of the IEA as well as with all new exporting countries which may develop.

Environment: Coal using units above a certain capacity have to be approved by the authorities. The pollution components which are considered are sulphur dioxide, nitrogen oxide and total quantity of suspended particulates.

RD&D: Coal utilization and combustion will receive funding of D.Kr. 24.8 million for the 1980-82 period.

Summary of Specific Issues

Parliament has decided that the government should recommend that power stations diversify their imports geographically. The implementation of this recommendation should be seen together with further development of long-term commitments and continuing efforts to maintain the adequacy of unloading and transhipment capacities.

The penetration of coal for electricity generation, supported by adequate reception facilities, has reached a high level. It now appears timely to support a major effort to stimulate penetration of coal in industry for large, medium and even small process heat consumers, where coal can be used on a competitive basis; the RD&D programme is serving this end. This effort should also be eased by the current infrastructure in place and the location of most large industries and DH plants close to the coastline, and by inland distribution of coal from public ports, allowing for transportation at reasonable costs.

DENMARK

Key Solid Fuel and Energy Indicators
(Mtoe)

	1973	1979	1980	1990	Average Annual Growth Rates (%) 1973-1980	1980-1990
Production						
Total Energy Production	0.1	0.7	0.6	7.5	33.7	29.5
Solid Fuel Production	0	0.3	0.3	0.3	n.a.	1.4
Solid Fuels as % of Total	0	37.2	46.0	4.0		
Trade						
Coal Exports	0	0	0	0	0.0	0.0
Coal Imports	2.0	4.6	6.1	7.9	17.0	2.6
Requirements						
Total Primary Energy Requirements	19.5	20.8	19.2	19.8	- 0.2	0.3
Solid Fuel Requirement	2.1	4.6	6.1	8.2	16.9	2.9
Solid Fuels as % of Total	10.5	22.1	31.9	41.4		
Sectoral Consumption of Solid Fuels and Comparison to Oil						
Solid Fuel Use						
Industry [1]	0.3	0.6	0.5	0.9	9.9	5.9
Iron and Steel	0	0	0	n.a.	6.5	n.a.
Other	0.2	0.5	0.5	n.a.	10.2	n.a.
Residential and Commercial [2]	0.1	0.3	0.3	0.6	13.9	6.7
Electric Generation [3]	1.7	3.7	5.5	6.8	18.6	2.1
Solid Fuel Share (%)						
Industry [1]	7.3	17.3	16.7	25.0		
Iron and Steel	9.9	13.3	13.4	n.a.		
Other	8.4	19.8	18.7	n.a.		
Residential and Commercial [2]	1.5	2.9	4.0	7.0		
Electric Generation [3]	36.8	65.2	82.1	82.9		
Oil Share (%)						
Industry [1]	82.6	66.9	66.0	38.9		
Iron and Steel	85.7	60.4	56.9	n.a.		
Other	82.2	67.2	67.6	n.a.		
Residential and Commercial [2]	85.8	80.5	77.0	34.9		
Electric Generation [3]	63.0	34.7	17.8	12.2		

1. Excludes non-energy use.
2. Includes Public and Agricultural Use
3. Represents fuel input to electric generation

GERMANY

Coal Use: For electricity generation the current ten-year contract (ending in 1987) between the coal mining industry and the electricity generating companies provides for supplies of indigenous coal totalling 33 million tons per year. A new contract between the same partners extends the present arrangements until 1995, assuring the sale of some 640 million tons of coal in the period 1981 to 1995. Broken down into periods, the agreement involves the following indigenous supplies:

1981-1985:	38 million tons per year
1986-90:	43 million tons per year
1991-95:	46 million tons per year

Public and industrial utilities will enter into individual buying commitments with indigenous hard coal producers as a condition for certain import quotas for additional coal. In the past, indigenous coal was not competitive either in power plants or in industry, and the sale of coal was guaranteed largely by laws on electricity generation and also by means of import regulations. The price of indigenous production was brought into line with oil prices through a subsidy scheme. To ensure the utilization of coal for electricity generation, it is necessary to continue the Coal Utilization Fund, which is paid by the electricity user in the form of a coal levy (at present 4.2% of the electricity price).

The specific aims of financial aid from this compensatory levy are:

- compensation to offset the price difference between domestic coal and heavy heating oil (on average, for about 22 million tce/year);
- compensation to cover the difference between domestic coal costs and the average price of imported coal delivered to the power station (on average this constitutes about 11 million tce/year; thus all power station operators receive this amount of coal at the same price, towards which they must pay DM 6/tce);
- investment grants for the construction of new coal-fired power stations (DM 180/kw) and new coal-fired heat and power stations (maximum aid DM 480/kw) and for the conversion of oil-fired heat and power stations;
- special aids for the sale of coal of lower burning quality (low grade coal, low volatile coal) and for sales outside the Ruhr area.

For volumes exceeding 33 million tce/year, power station operators receive coupons entitling them to imported coal from third (non-Common market) countries on the basis of a ratio of 2:1 from 1981 to 1987 and 1:1 after 1988.

It has been government policy for a number of years not to issue licences for new oil- and gas-fired power plants. The conclusion of the new agreement between the electricity and coal industries, the liberalization of import quotas and the rising crude oil prices, all tend to favour a further reduction of oil use in power plants. According to steps taken by the federal government, the electric utility industry has undertaken not to resort to oil-fired power plants unless unavoidable. In fact the public electric utilities now only use oil-fired capacity for peakload and as reserve capacity.

Since mid-1979, industry has started to reconsider the use of coal. But conversion to coal requires considerable investment and the need to meet very strict environmental requirements. In order to create the incentives needed, the increase in the coal import quota allows industrial consumers to obtain coal from third countries outside the Common market for the years until 1984 on condition that such imports displace oil. In 1980 total imports of coal increased by 16% compared to 1979. In 1980, consumption of heavy fuel oil in industry dropped by 14% compared to 1979 but coal consumption increased by 5%, while total primary energy consumption fell by 5%. An agreement on co-operation between the individual sectors of industry (industrial power plant operators, federation of German industries and association of municipal entreprises) has been reached; it will aim at encouraging and supporting the utilization of exploitable waste heat potential and supplying and making use of industrial waste heat.

Energy Prices: The federal government remains committed to a policy of free energy price formation according to the world market. As a basic principle, prices are not regulated, although there are two exceptions to this rule:

— domestic coal prices are subsidized, through the special levy on electricity, in order to balance the cost advantage of imported coal;

— electricity tariffs to households and other small-scale consumers are subject to approval by the Länder governments, at the request of local utilities.

Coal Production: Hard coal and lignite mining is governed by the federal mining law (Bundesberggesetz), which superseded the mining laws of the Länder from the beginning of 1982. They assign the holders of a mining permit a legal claim to cession by the land-owners of the area required for mining operation. The area required for exclusively hard coal mining is restricted basically to the necessary space for the pit and ancillary installations on the surface surrounding the shaft and for dumping the mining waste and washery rejects. Considerable constraints to hard coal moving stem from the damage done to overlying land and buildings, which occurs as a result of underground mining and for which mining companies are obliged to pay compensation.

Special legal arrangements beyond the mining laws take account of lignite mining requirements (North Rhine/Westphalia) through a special section of the Provincial Planning Act which superseded the 1950 Lignite Mining Act. On the basis of long term planning and co-operation of all interested parties (mining operators, regional and local authorities, mining supervisory authorities, etc.) the lignite mining industry is granted the required access to these deposits with due consideration for the interests of the population affected. A major provision of this legislation is mandatory land reclamation after mining.

Possibilities of increased lignite mining are strictly limited owing to the concentration of deposits in the densely populated Rhineland (90% of present production) and to the extensive surface required for production. Hard coal production development possibilities are also limited. For the time being the hard coal industry is aiming at a stabilization of present production capacity at around 90 million tons per year followed by an increase to around 95 million tons per year from the 1990's, if this is warranted by overall economic development. The proposed development is also subject to other considerations, including proof of sufficient recoverable deposits through increased exploration; balanced production and earnings trends allowing the industry to finance investments in new pits; and long-term availability of sufficient labour.

Transportation: Facilities for coal transport which include railways, inland waterways, road systems and ports are adequate, and could cope with further traffic increase resulting from an expansion of coal demand.

Trade: Expanded imports of coal are now possible for utilities and industrial consumers. The 1980 Coal Tariff Quota Law introduced new provisions for imports from non-European Economic Community (EEC) countries. It will apply until the end of 1995 and provides for growing import quotas which the federal government can increase under certain conditions for various users and will mainly aim at favouring substitution away from oil. The quotas, currently considered as adequate to cover future import requirements, amount to:

for 1985: about 23 million tons (can be increased to about 31 million tons);

for 1990: about 31 million tons (can be increased to about 43 million tons);

for 1995: about 39 million tons (can be increased to about 55 million tons).

Of these quantities, the following amounts will be available for power stations:

in 1985: about 9 million tons (can be increased to about 11 million tons);

in 1990: about 13 million tons (can be increased to about 17 million tons);

in 1995: about 17 million tons (can be increased to about 23 million tons).

Until the end of 1983, industries wishing to take advantage of the quotas are obliged to submit proof to the effect that imported coal will actually displace oil or gas.

Imported coal is increasingly being supplied under long-term contracts or through participation in foreign coal mines. However, substantial losses of coal supplies from Poland have given rise to some concern, and these losses have had to be partially offset, in 1980-81, on spot markets, with some difficulty and at higher cost.

Environment: The increasing difficulties encountered in the densely populated mining and industrial areas against the building up of coal tips is an issue.

Plans to continue and increase coal and lignite use for power generation will require more stringent clean air requirements, which will be met by the development of more efficient dedusting and flue gas desulphurization methods (including waste disposal resulting from these methods), as well as by the development of low polluting combustion methods, such as fluidized bed combustion and partial fuel gasification. The development of these methods is supported by financial aid from the government through the Energy Research and Energy Technology Programme.

RD&D: For power station technology, three possible methods of advanced coal combustion are currently being developed:

— combined cycle with atmospheric fluidized bed combustion and gas turbine (MKV — Modell Kraftwerk Völklingen-Saarbergwerke AG — demonstration power plant of 230 MW put into operation in late 1982);

— pressurized fluidized bed combustion (IEA Research programme in the United Kingdom - 85 MW);

— VEW (Vereinigte Elektrizitatswerke Westfalen A.G.) coal conversion process.

These processes aim at making costly desulphurization unnecessary and making hard coal use compatible for running gas turbines efficiently. Atmospheric fluidized bed combustion can be applied in district heating, local heating plants and industrial steam generators. Prototypes of various sizes are under construction or planned.

Since 1974 great efforts have been undertaken to develop modern technologies of coal gasification and coal liquefaction. Eight pilot plants for coal gasification are operating (1-10 tons of coal per hour). Two pilot plants for coal liquefaction were commissioned in 1981 (with a coal throughput of 200 tons and 6 tons per day).

In October 1981, the government decided on further steps to implement the Coal Conversion Program of January 1980. For coal gasification demonstration projects the following assistance is available: an investment grant not to exceed 40% (in justified cases not to exceed 50%); if German hard coal is used, a price differential is paid for a maximum of DM 60 per ton for five years to bring coal cost in coal gasification projects to levels competitive with other countries. Feasibility studies have been submitted for seven gasification projects of which the most promising will be assisted. Federal funds available between 1982 and 1985 for this purpose total DM 1 billion.

A government decision on similar financial assistance to coal liquefaction projects is to be taken in the second half of 1982.

Summary of Specific Issues

In addition to the recent relaxation of import quotas, a flexible implementation of the coal import policy, together with streamlining administrative procedures for conversion and financial aid, would help to stimulate industries to switch from oil to coal.

Domestic production of both coal and lignite faces the major obstacle of land availability, even to maintain current levels of production. As regards hard coal production, important subsidies are already involved to maintain the present level of production and any increase would therefore be difficult. The relevant issue is to find an acceptable solution for a secure, adequate and economically bearable solid fuel supply (domestic and imported coal) in an environmentally acceptable way. The high cost of environmental protection for coal, together with financial constraints (oil-fired installation not yet written off in full) and in certain instances lack of infrastructure (storage facilities, etc.), constitutes a major constraint to fuel switching, particularly in industry. The small share of energy cost in many final products is also an obstacle to converting to coal.

GERMANY

Key Solid Fuel and Energy Indicators
(Mtoe)

	1973	1979	1980	1990	Average Annual Growth Rates (%) 1973-1980	1980-1990
Production						
Total Energy Production	121.1	124.4	124.0	150.0	0.3	1.9
Solid Fuel Production	92.8	89.6	90.3	89.0	- 0.4	- 0.1
Solid Fuels as % of Total	76.7	72.0	72.9	59.3		
Trade						
Coal Exports	17.4	19.1	14.6	7.0	2.5	- 7.1
Coal Imports	7.0	7.0	8.1	13.0	2.0	4.8
Requirements						
Total Primary Energy Requirements	266.4	285.0	272.2	306.0	0.3	1.2
Solid Fuel Requirement	84.0	81.5	83.5	95.0	- 0.1	1.3
Solid Fuels as % of Total	31.5	28.6	30.7	31.0		
Sectoral Consumption of Solid Fuels and Comparison to Oil						
Solid Fuel Use						
Industry [1]	18.6	16.4	17.3	16.0	- 1.1	- 0.8
Iron and Steel	14.1	11.8	12.0	n.a.	- 2.2	n.a.
Other	2.8	3.1	3.5	n.a.	3.2	n.a.
Residential and Commercial [2]	10.1	6.1	5.9	4.0	- 7.4	- 3.9
Electric Generation [3]	47.3	53.0	54.0	60.0	1.9	1.1
Solid Fuel Share (%)						
Industry [1]	22.7	21.0	23.5	23.5		
Iron and Steel	53.9	58.1	62.9	n.a.		
Other	8.6	9.9	11.6	n.a.		
Residential and Commercial [2]	12.8	7.2	7.6	4.7		
Electric Generation [3]	69.9	61.6	63.5	52.2		
Oil Share (%)						
Industry [1]	45.9	41.1	38.8	16.2		
Iron and Steel	17.1	11.6	7.5	n.a.		
Other	58.7	45.6	44.2	n.a.		
Residential and Commercial [2]	64.7	59.8	54.2	45.9		
Electric Generation [3]	9.9	6.7	6.3	2.6		

1. Excludes non-energy use.
2. Includes Public and Agricultural Use
3. Represents fuel input to electric generation

GREECE

Coal Use: Major development of domestic lignite production is planned to support electricity generation. The development programme of the Public Power Corporation (PPC) is based primarily on lignite (3300 MW) and approximately 2900 MW of hydro power; no new oil-fired electricity generating capacity is planned for continental Greece.

The planning for the introduction of imported coal in energy-intensive industries is gradually taking shape. A new law 1116/81 on incentives for industry, providing direct grants up to 40% for energy conservation and substitution of oil and electricity, is in force and several applications for grants have been approved. By the end of 1982 about 1.5 million tons of coal will be used in the cement industry, replacing about 0.9 million tons of fuel oil.

Energy Prices: Fuel oil prices have been gradually brought into line with world market prices. There are no excise taxes on heavy fuel oil.

Coal Production: An almost tripling of the current level (4.6 Mtce in 1980) of lignite production is foreseen for 1990.

Transportation: Coal imports will require construction of reception facilities near major industrial users.

Trade: The supply of 1.4-1.6 million tons of coal is being considered in parallel with the reconversion of energy-intensive industries (such as cement plants) from oil to coal.

RD&D: Of the US$ 4.6 million energy R&D budget, 30% is devoted to solid fuel research, in particular more efficient utilization of indigenous energy resources.

Summary of Specific Issues

The change observed in the industrial sector under the provision of Law 1116/81 is encouraging, and an increase in the amount of funding (US$ 6.83 million in 1981) in coming years should be considered to reinforce and expand results.

A downward revision of future electricity demand has postponed plans for the construction of power stations fired with imported coal as anticipated development of lignite deposits would match requirements for electricity generation.

GREECE

Key Solid Fuel and Energy Indicators
(Mtoe)

	1973	1979	1980	1990	Average Annual Growth Rates (%) 1973-1980	1980-1990
Production						
Total Energy Production	2.4	4.2	4.1	10.5	7.9	9.9
Solid Fuel Production	1.8	3.3	3.2	8.8	8.5	10.8
Solid Fuels as % of Total	75.8	78.1	78.4	83.8		
Trade						
Coal Exports	0	0	0	0	n.a.	n.a.
Coal Imports	0.5	0.5	0.4	1.6	- 2.9	15.0
Requirements						
Total Primary Energy Requirements	12.3	16.2	16.2	25.7	4.0	4.7
Solid Fuel Requirement	2.3	3.7	3.5	10.4	6.4	11.4
Solid Fuels as % of Total	18.6	23.0	21.8	40.5		
Sectoral Consumption of Solid Fuels and Comparison to Oil						
Solid Fuel Use						
Industry [1]	0.6	0.6	0.5	1.7	- 2.1	12.9
Iron and Steel	0.4	0.3	0.3	n.a.	- 3.8	n.a.
Other	0.1	0.1	0.1	n.a.	2.7	n.a.
Residential and Commercial [2]	0.1	0.1	0.1	n.a.	- 0.5	n.a.
Electric Generation [2]	1.5	2.9	2.8	8.4	9.4	11.4
Solid Fuel Share (%)						
Industry [1]	17.4	13.1	12.0	22.4		
Iron and Steel	42.8	58.1	55.5	n.a.		
Other	3.7	3.1	2.5	n.a.		
Residential and Commercial [2]	2.1	2.4	2.0	n.a.		
Electric Generation [3]	39.3	50.6	48.6	83.2		
Oil Share (%)						
Industry [1]	63.7	68.0	68.4	55.3		
Iron and Steel	26.9	29.3	30.0	n.a.		
Other	82.2	77.6	77.8	n.a.		
Residential and Commercial [2]	80.7	71.3	69.0	n.a.		
Electric Generation [3]	45.7	33.2	36.4	4.0		

1. Excludes non-energy use.
2. Includes Public and Agricultural Use
3. Represents fuel input to electric generation

IRELAND

Coal Use: Present policy is that all new base load electricity generation plants will burn solid fuels. Coal use in this sector is presently very small (40,000 tons for a 15 MW station). The possibility to convert modern oil-fired boilers to coal has been considered but the rate of growth of electricity demand in recent years has required that emphasis be placed on constructing new coal-fired units. The Electricity State Board is planning some 900 MW capacity for the current decade of which 600 MW will be in service in 1986. Grants are available for switching from oil to coal in industry, and a number of large industrial users of oil (cement, sugar, dairy industries) are converting from oil to coal. A grant scheme to support the change from oil-fired central heating systems to the use of solid fuels has recently been abandoned, as it was felt that the economics of fuel switching are favourable enough that such support was no longer required.

Energy Prices: A general price control system is in place and maximum retail prices are set for a number of essential commodities which include all fuels. It is agreed that the prices set for oil products should compensate for actual costs. Taxation has been organized so as to impose penalties on oil consumption. There is an excise duty on fuel oil of IR* £20 per tonne and Value Added Taxation (VAT) on petrol and diesel fuel, with exemption from taxes for other fuels (no VAT on peat, coal, gas or electricity).

Coal Production: Indigenous solid fuel resources and production are small, and exploitation is concentrated on peat mainly through the State Peat Board. A current issue is the progressive exhaustion of large surfaces of bogland for future exploitation. A State grant scheme was introduced in July 1981 for the development of privately-owned bogs.

Transportation: The amount of inland transportation capacity required is not substantial. The major problem is port facilities to handle increased imports. Facilities are being provided to enable the electricity authority to handle increased volumes of coal and a deep harbour, capable of handling fully-laden bulk carriers of up to 160,000-170,000 dwt, is being built in connection with the Moneypoint power plant. Initially this terminal will be equipped only for unloading and is not linked to the national railway grid. However, this facility may eventually be used for handling coal for industry.

RD&D: Plans are proceeding for building a fluidized bed unit for electricity generation. This would make possible the exploitation of domestic low quality coals on a significant scale. A design competition has been launched on multi-solid fuel domestic stoves.

---

* US$ 1 = 0.64 IR £ as at 31 December 1981.

Summary of Specific Issues

Diversification of supply sources for future coal imports is a necessity. The reliance on Poland as the main supplier created major problems in 1980. However, this overdependence has ceased and imports are now more diversified.

Incentives to accelerate fuel switching in both industrial and residential/commercial sectors are needed on a continued basis to maintain the momentum in increasing the share of coal use.

IRELAND

Key Solid Fuel and Energy Indicators
(Mtoe)

	1973	1979	1980	1990	Average Annual Growth Rates (%) 1973-1980	1980-1990
Production						
Total Energy Production	1.2	1.7	2.1	3.5	8.1	5.1
Solid Fuel Production	1.1	0.9	1.1	1.7	0.3	4.8
Solid Fuels as % of Total	86.6	54.9	51.1	49.6		
Trade						
Coal Exports	0.1	0.1	0	0	- 9.2	0.0
Coal Imports	0.6	0.9	0.8	3.3	5.5	14.7
Requirements						
Total Primary Energy Requirements	7.4	9.0	8.6	12.9	2.2	4.1
Solid Fuel Requirement	1.6	2.0	1.9	5.0	2.6	10.3
Solid Fuels as % of Total	21.4	22.1	21.9	39.1		
Sectoral Consumption of Solid Fuels and Comparison to Oil						
Solid Fuel Use						
Industry [1]	0	0	0	0.5	n.a.	n.a.
Iron and Steel	0	0	0	0	n.a.	n.a.
Other	0	0	0	0.5	n.a.	n.a.
Residential and Commercial [2]	0.9	1.4	1.3	1.7	5.5	2.6
Electric Generation [3]	0.6	0.6	0.6	2.9	0.0	17.2
Solid Fuel Share (%)						
Industry [1]	3.2	1.1	0.3	15.1		
Iron and Steel	19.0	52.0	26.5	n.a.		
Other	3.3	0	0	n.a.		
Residential and Commercial [2]	35.3	46.1	46.8	48.1		
Electric Generation [3]	34.3	21.6	21.8	66.2		
Oil Share (%)						
Industry [1]	80.8	69.5	65.7	58.9		
Iron and Steel	68.0	36.4	67.1	n.a.		
Other	79.7	83.3	81.9	n.a.		
Residential and Commercial [2]	47.9	36.1	33.8	25.2		
Electric Generation [3]	56.9	60.2	53.0	14.9		

1. Excludes non-energy use.
2. Includes Public and Agricultural Use
3. Represents fuel input to electric generation

ITALY

Coal Use: Increased coal use is one of the major goals of the new National Energy Plan (NEP) for 1980-1990, adopted by government and Parliament at the end of 1981. The Plan foresees coal consumption for electricity generation of about 32 Mtce by 1990, and this would amount to 32% of fuel input for the utility sector compared with just above 10% in 1980. The exclusion of oil products from new base load power stations has been confirmed, and a conversion programme is under way. By 1985, 2.6 GW should be converted from oil to coal, resulting in 7.3 GW of coal-fired plant in 1985, compared with 4.7 GW in 1980. By 1990 this capacity should increase to 19.7 GW.

In the industrial sector, the increase of coal use between 1980 and 1990 is projected to be less than 3 Mtoe. Although the cement and building materials industries are promising areas for coal use, actual switching to coal in industry in general is proceeding very slowly. The Restructuring and Reconversion of Industry Law (675/77) contains some incentives which may be used to foster coal use in industry. In the residential sector major obstacles to coal use are the small size of average consumption units and pollution considerations. A proposal (Draft Bill 2383) to allow financial incentives to local and regional authorities for electricity generated from sources other than hydrocarbons has been approved by the Senate but it still awaits final approval by the Chamber of Deputies.

Energy Prices: Fuel oil prices are monitored which means that the government observes price changes to assure they reflect costs and are in line with the world market. Gas prices to industry and for electricity generation are linked to those of fuel oil.

Coal Production: The only coal reserve is located in Sardinia: quantity is low and sulphur content is high. A plan has been developed to produce coal for electricity generation, either through gasification or direct use. The decision to resume mining appears to be motivated more by regional and security concerns than by economic considerations and the bulk of increase in coal consumption will be met only through imports.

Transportation: Most imported coal comes by ship and Italian ports are in most cases not equipped for the reception of large quantities of steam coal arriving in large coal carriers. The ability to expand the role of coal will depend critically on the development of coal handling infrastructure, particularly the development of ports and inland rail systems. Three major coal ports are foreseen in the NEP: Trieste/Porto Levante, Vado Ligure and Milazzo/Gioia Tauro. The Plan estimates a total cost of 1200 billion Lire, to be equally shared between port development and inland distribution facilities. Timely implementation of these very large interrelated projects will require close co-operation between the Ministry of Industry, the Ministry of Public Works and the interested parties in trade. Development of such facilities could also provide an opportunity for Italy to become a transit trader in new coal markets. A feasibility study is presently being conducted for the coal terminal project in Trieste: a two stage development plan is foreseen with the first stage in operation by 1985 (5 million tons/year) and the second by 1990 (another 9 million tons/year). Access by vessels up to 150,000 dwt would be possible after completion of the first stage. This terminal could also be used for coal imports into Austria. Decisions about the other coal terminals are still pending.

Trade: Italy is already importing coal from major world coal suppliers. State-owned companies are active in the international market, through negotiation of medium- and long-term contracts and participation in mining industries abroad, in order to diversify supply sources. Private companies are also active in this sector.

RD&D: It is foreseen that two joint processes will be utilized in developing domestic coal reserves: gas-from-coal and methanol from medium heating value gas obtained from coal. The financial aspects of these projects are currently being examined.

Summary of Specific Issues

The implementation of the recently approved NEP through legislative measures, followed by public and private initiatives, is a critical issue. The most important and urgent points which need to be defined are:

- location of coal centres suitable to receive large-scale coal carriers and inland storage capacity for distribution to small consumers;
- completion of site locations for early construction of new coal-fired power plants for a priority capacity of 6 GW;
- approval of DDL 2383, currently before Parliament which, **inter alia,** provides incentives for construction of non-oil-fired power plants to regions and local communities.

The development of coal to the extent indicated by the NEP is a very complex matter where many decisions in different sectors of the economy have to be taken in parallel and in a well-planned way. Only a clear and strong policy can overcome the difficulties throughout the negotiation process with regions and communities, in order to assure that the target can be reached. Urgent and specific measures are required in order to build up the necessary infrastructure and create a favourable climate for national and foreign investment.

ITALY

Key Solid Fuel and Energy Indicators
(Mtoe)

	1973	1979	1980	1990	Average Annual Growth Rates (%) 1973-1980	1980-1990
Production						
Total Energy Production	26.0	25.5	25.4	32.4	- 0.3	2.5
Solid Fuel Production	1.8	1.1	1.3	2.0	- 4.8	4.3
Solid Fuels as % of Total	7.1	4.4	5.2	6.2		
Trade						
Coal Exports	0.4	0.5	0.5	0	2.5	n.a.
Coal Imports	8.8	10.7	12.6	32.0	5.3	9.7
Requirements						
Total Primary Energy Requirements	132.9	143.5	142.1	179.9	0.9	2.4
Solid Fuel Requirement	10.3	11.6	13.2	32.0	3.7	9.9
Solid Fuels as % of Total	7.7	8.1	9.3	18.9		
Sectoral Consumption of Solid Fuels and Comparison to Oil						
Solid Fuel Use						
Industry [1]	4.6	4.5	4.7	7.4	0.3	4.7
Iron and Steel	3.5	3.5	3.6	n.a.	0.2	n.a.
Other	0.9	0.8	0.9	n.a.	0.4	n.a.
Residential and Commercial [2]	1.6	0.9	0.8	1.3	- 8.9	4.8
Electric Generation [3]	1.6	3.0	5.1	22.4	18.2	15.9
Solid Fuel Share (%)						
Industry [1]	9.2	9.5	10.8	16.8		
Iron and Steel	45.7	41.2	42.6	n.a.		
Other	4.2	3.6	4.2	n.a.		
Residential and Commercial [2]	4.9	2.5	2.3	3.3		
Electric Generation [3]	4.8	7.6	12.2	31.7		
Oil Share (%)						
Industry [1]	59.1	49.4	44.3	21.8		
Iron and Steel	14.2	13.4	12.8	n.a.		
Other	59.4	49.6	47.0	n.a.		
Residential and Commercial [2]	71.2	59.4	56.7	33.1		
Electric Generation [3]	61.3	56.6	54.9	33.7		

1. Excludes non-energy use.
2. Includes Public and Agricultural Use
3. Represents fuel input to electric generation

JAPAN

Coal Use: No oil-fired power plants will be authorized, in principle, except those already under construction or planned. The government is determined to guide the utilities away from oil, which should be confined in principle to middle- and peak-load demand. Utilities have also been requested to check the possibility of converting existing oil-fired plants to non-oil fuels including coal. Measures adopted by government to increase the use of coal vary, and include financial assistance to utilities and industry for fuel switching, introduction of alternative energy sources and the development of combustion techniques which reduce impact on the environment.

For instance, the Japan Development Bank provides concessional financing for the construction of coal-fired power plants. Low interest loans are also provided to convert boilers to coal and for the construction of coal centres. Subsidies are provided for the construction of coal-fired plants and for the conversion of existing plants to coal firing. To ease electric power plant siting, the government has introduced a grant system whereby grants are provided to prefectures (provinces) supplying electricity to others, in order to meet the expenses of projects; to industrialize areas where a power plant is located; and to modernize the industries already existing in these areas. A tax credit system has been established for the promotion of energy policy measures including the increased use of coal; it is applicable since FY 81 and valid for three years. The tax credit can take the form of a 30% special depreciation or a 7% tax reduction for the first year of installation of new equipment. The switch from oil to coal is particularly sizeable in the cement industry where the share of coal has increased from 21% of total energy in April 1980 to 77.5% in March 1981.

Energy Prices: Prices of oil products are not controlled. Electricity tariffs and gas prices to consumers are regulated and determined on the basis of the costs sustained by the industry.

Coal Production: The government is supporting domestic production by providing various incentives. The current production level is around 18 million tons/year. Major constraints to increasing coal production are the high costs of mining, labour availability, difficulties in securing safety and limitations in the introduction of modern mining systems.

Through the New Energy Development Organization (NEDO) established in October, 1980, the government also provides financial assistance for exploration and development of overseas coal mines through loans for up to 70% of funds needed for exploration, and loan guarantees for the development of mines.

Transportation: Infrastructure for receiving coking coal is already handling 50-60 million tons per year. The expected sharp increase in steam coal imports requires the development of reception facilities for coal carriers.

There are three ways to handle imported coal:

— unloading and distributing via a coal centre;
— unloading at a public wharf and distributing to users;
— directly unloading at a private wharf of a user.

A coal centre with the capacity of handling 1.5 million tons of coal per year has been put into operation, and six ports have started to operate as receiving terminals of imported steam coal so as to meet increasing demand. Other construction plans of coal centres are to be carried out stage by stage, while improvement projects are now under way in three ports. The government is providing assistance for these improvements as well as for the construction of port facilities through the "Energy Port Supporting Scheme". This scheme is managed by the Ministry of Transport and provides financial participation for building breakwaters (25%) and channels, anchorages and basins (25% up to 13 metres depth; deeper than 13 metres, decreasing financial participation with a miminimum of 5%). The port authority contributes the same amount. These aids are applied to energy ports, contributing to diversification and increased stability of energy supplies, and they can cover coal ports, stockpiling for oil and reception facilities for LNG. MITI is supplying soft loans for 40% of the cost of equipment of coal centres. The government is also promoting the development of adequate coal carrier capacity.

Trade: The government considers that in order to cope with a substantial increase in coal demand it is necessary to establish a secure coal supply chain relying upon the development of coal mines in coal-producing countries, improvement of infrastructure and development of coal carrier capacity. As for coal imports, the total amount of contracted quantities in 1985 and 1990 cannot be estimated by the government. However, it seems that almost all coal imports planned for 1985 have already been assured through contracts or development participation, but that a substantial share of the coal imports planned for 1990 remains to be contracted.

In order to promote the development of overseas coal mines various kinds of assistance are provided by NEDO (subsidies for surveys on potentiality of overseas coal mines development, survey on geological structures abroad, etc.).

Environment: Environmental considerations have great importance because of high population density and scarcity of land. Some policies are being developed aimed at reducing emissions and managing ash disposal. Some opposition from local residents against coal-fired plants still remains a problem.

RD&D: Significant efforts are being devoted, with government subsidies, to the development of non-polluting combustion techniques. Many different types of equipment are being tested with the objective of establishing reliable technologies for the expanded future use of coal. Japan participates in the IEA co-operative R&D programme on atmospheric fluidized bed combustion and in the IEA programme on coal-oil mixture technology for utilization in large boilers. A survey at the national level has been undertaken to find adequate sites for large-scale concentrated ash disposal yards; preliminary planning for ash disposal yards has also been carried out.

In the coal liquefaction area, Japanese industry is participating in the Exxon Donor Solvent pilot plant. Four Japanese liquefaction processes are in the early R&D stages. Low and high calorific value gas from coal technology is also under development: one pilot plant for low-calorific gasification with 40 tons/day capacity is completed and currently in operation, while a high-calorific gasification plant with 7000 cubic metres/day capacity is under construction.

Summary of Specific Issues

The coal development programme is very strong and ambitious. In addition to maintaining the current level of domestic production, it foresees nearly a doubling of imports from about 76 Mtce in 1980 to 141 Mtce in 1990.

As for coal receiving terminals, the construction of coal centres is underway in addition to other measures. It is critical that these reception facilities be developed in a timely manner to meet an increase in coal imports.

The large increase in projected coal demand (33% for electricity generation and 66% for industry) requires considerable effort to secure supplies under reasonable conditions; investment in producing areas might be a necessity.

Timely development of new electricity generating capacity and the very large increase in industry demand could be critical when considering the complexity of environmental constraints, due to specific local conditions such as high density of population and scarce land availability.

In industry, major developments may be subject to the results of improved combustion technologies, presently being pursued vigorously with substantial government aid. Some of these technologies, mainly centred on emission control, have not yet reached the demonstration-conversion stage. Another critical issue is the disposal and treatment for large quantities of ash and sludge for which preliminary studies are under way. The time schedule for the implementation of both new combustion technologies and waste management is uncertain.

JAPAN

Key Solid Fuel and Energy Indicators*

(Mtoe)

	1973	1979	1980	1990	Average Annual Growth Rates (%) 1973-1980	1980-1990
Production						
Total Energy Production	42.1	53.8	59.0	131.0	4.9	8.3
Solid Fuel Production	18.8	13.1	13.2	13.2	- 4.8	0
Solid Fuels as % of Total	44.6	24.3	22.4	10.1		
Trade						
Coal Exports	0.4	1.4	1.4	—	20.0	n.a.
Coal Imports	42.4	43.4	53.2	98.6	3.3	6.4
Requirements						
Total Primary Energy Requirements	340.6	376.8	372.7	518.8	1.3	3.4
Solid Fuel Requirements	59.8	55.8	66.4	111.8	1.5	5.4
Solid Fuels as % of Total	17.6	14.8	17.8	21.5		
Sectoral Consumption of Solid Fuels and Comparison to Oil						
Solid Fuel Use						
Industry [1]	32.7	29.9	36.4	68.1	1.5	6.5
Iron and Steel	30.8	26.9	28.7	n.a.	- 1.0	n.a.
Other	1.9	2.5	7.1	n.a.	20.6	n.a.
Residential and Commercial [2]	4.9	3.0	3.7	0.4	- 4.1	- 20.0
Electric Generation [3]	12.2	13.6	14.6	35.4	2.7	9.3
Solid Fuel Share (%)						
Industry [1]	23.3	19.9	25.3	35.4		
Iron and Steel	63.4	66.0	71.9	n.a.		
Other	3.6	4.0	11.4	n.a.		
Residential and Commercial [2]	9.1	4.8	6.1	0.5		
Electric Generation [3]	11.5	10.9	11.7	17.2		
Oil Share (%)						
Industry [1]	57.2	59.3	53.8	40.5		
Iron and Steel	23.8	17.8	12.3	n.a.		
Other	67.9	65.6	58.3	n.a.		
Residential and Commercial [2]	63.8	59.8	56.6	39.2		
Electric Generation [3]	67.6	48.2	41.1	18.1		

1. Excludes non-energy use.
2. Includes Public and Agricultural Use
3. Represents fuel input to electric generation

* The new Long Term Energy Supply and Demand Outlook was published by the Japanese authorities on 21 April 1982. The new values have been incorporated in the forecasts of this report.

LUXEMBOURG

Coal Use: Total coal use diminished over the 1973-78 period in both percentage and absolute terms due to economic recession. But since 1979 it has increased and displaced oil products. The only electricity production in Luxembourg is the auto-production of the iron and steel industry and more than two-thirds of the electricity consumed is imported. In 1980 energy demand fell by 6.5% from 1979 levels, but the oil reduction (17%) was even more significant and was partly the result of substitution of coal for oil in the industry sector, in particular in the iron and steel industry.

Adequate future coal supplies have been secured through the iron and steel industry's participation in German mines.

A projected coal-fired power plant has been postponed because the Belgian partner to the scheme currently has excess production capacity due to stagnation of consumption and because the provision of water for cooling purposes raises considerable problems for the proposed site.

Energy Prices: Almost all energy product prices have to be approved by the government. Considerable price increases have taken place in the last years, in particular for gas and oil products; however, as a consequence of the general tax policy with a high level of direct taxes and low level of indirect taxes, some prices, especially petrol and natural gas prices, remain at a lower level than in neighbouring countries.

Trade: Luxembourg relies exclusively on imported coal and security of supplies can be considered assured until 1995 through the participation of the iron and steel industry in German coal mines.

LUXEMBOURG

Key Solid Fuel and Energy Indicators
(Mtoe)

	1973	1979	1980	1990	Average Annual Growth Rates (%) 1973-1980	1980-1990
Production						
Total Energy Production	0.3	0	0.1	0	n.a.	n.a.
Solid Fuel Production	0	0	0	0	n.a.	n.a.
Solid Fuels as % of Total						
Trade						
Coal Exports	0	0	0	0	n.a.	n.a.
Coal Imports	2.4	1.9	1.8	2.2	- 3.9	1.9
Requirements						
Total Primary Energy Requirements	4.8	3.8	3.6	4.7	- 3.7	2.6
Solid Fuel Requirement	2.4	1.8	1.8	2.2	- 4.0	2.0
Solid Fuels as % of Total	50.5	47.6	49.7	46.8		
Sectoral Consumption of Solid Fuels and Comparison to Oil						
Solid Fuel Use						
Industry [1]	2.1	1.6	1.5	2.0	- 4.3	2.7
Iron and Steel	2.1	1.6	1.5	n.a.	- 4.9	n.a.
Other	0	0	0.1	n.a.	23.2	n.a.
Residential and Commercial [2]	0	0	0	n.a.	- 4.6	n.a.
Electric Generation [3]	0.3	0.1	0.1	0.2	- 9.2	3.2
Solid Fuel Share (%)						
Industry [1]	64.8	67.5	70.3	64.5		
Iron and Steel	69.1	72.9	76.3	n.a.		
Other	12.5	29.1	43.0	n.a.		
Residential and Commercial [2]	6.0	3.9	4.1	n.a.		
Electric Generation [3]	40.4	42.0	47.0	68.7		
Oil Share (%)						
Industry [1]	24.4	14.5	8.3	19.4		
Iron and Steel	21.2	10.1	4.2	n.a.		
Other	59.9	48.1	28.1	n.a.		
Residential and Commercial [2]	78.5	67.3	62.1	n.a.		
Electric Generation [3]	15.3	9.0	4.9	0		

1. Excludes non-energy use.
2. Includes Public and Agricultural Use
3. Represents fuel input to electric generation

THE NETHERLANDS

Coal Use: The most important recent development in coal policy has been the completion and publication of the Coal Memorandum in February 1980. A main objective of this policy is to diversify Dutch energy supplies by increasing coal use to at least 40% of electricity generation by 2000. Oil use for electricity generation has increased substantially, to 5.1 Mtoe in 1980 compared with 1.6 Mtoe in 1973, while gas use has been reduced from 9.6 Mtoe in 1973 to 5.9 Mtoe in 1980. A relatively modest quantity of coal was used in 1980 (2.7 Mtce). New provisions for 1985 and 1990, established in December 1981, reflect a reversal of this trend: for 1985 electricity generation would absorb 4.4 Mtce of coal and by 1990 coal use is expected to increase to 9.9 Mtce. This would represent 52% of fuel inputs for electricity generation and reduce oil use to 4%. Prohibition of the construction of new oil- and gas-fired power plants is used by the government as an instrument to increase coal use in power generation. Even though there is an overcapacity in power plants amounting to about 25% of total capacity, new coal-fired plants have been built (a second 600 MW plant has recently started operation) but the excess capacity is expected to be absorbed more slowly than foreseen. A new 600 MW coal-fired power plant will be ordered in 1982.

The policy of introducing coal in electricity generation is focusing also on conversion of existing oil- and gas-fired capacity. The conversion of two units with a total capacity of 300 MW has recently been completed, and a programme is being outlined for converting existing power stations from oil or gas to coal (2000 MW out of a total of 10,000 MW oil/gas fired). The impact of such a programme on the temporary decommissioning of the power stations concerned, together with the economic aspects of the costs of the supplementary investments versus the possible evolution of the relative price of coal, are the main topics currently under consideration.

The attraction of coal-firing in industrial markets is believed to be low, despite substantially lower fuel prices, because of the high investment needed for coal-fired installations, stringent environmental standards and the general lack of space. The Investment Account Act provides for a premium of 15% on a permanent basis for all investments for coal burning and coal conversion projects in industry. Use of coal in combined heating power plants is currently being studied. In the longer term, new combustion technologies are seen as a promising solution to prevent environmental difficulties.

A Committee on CHP in industry published its report at the end of 1980. Until recently oil was used for CHP but now the policy is directed to conversion to coal. Until coal can be used, gas will be made available to move away from oil.

Industrial CHP is based on oil and gas, and DH schemes are based on oil, gas and waste heat. To stimulate CHP in industry and to maintain the possibility to use coal in the future, continued deliveries of gas will be made available in coming years.

A large-scale gasification project is planned at Eemshaven (Northern part of the Netherlands but final decision on whether to build this plant has been postponed until the end of 1982). This will produce low-calorific gas to be used for downgrading the high-calorific gas from the offshore deposits to standard (Groningen Gas) quality. Another use would be for chemical feedstock (methanol production). A decision should be taken in 1982 and, if positive, production is slated to start in 1985. The anticipated use of coal is estimated at 2 million tons/year. Detailed engineering work is also being carried out for the realization of a large-scale demonstration project for coal gasification: the start up of the plant is planned in 1985 and the produced gas will be delivered to a power station.

Energy Prices: Oil product prices are set by the Minister of Economic Affairs, after consultation with the oil industry, on the basis of Rotterdam spot prices and overall supply costs of the oil industry. The fuel oil price is in line with the world price. It is the intention of the government to price natural gas at a premium over relevant oil product prices. Despite rapid increase in gas prices in recent years, however, there still exist considerable differences between gas prices and the higher equivalent oil product prices especially in the residential sector and in the horticultural industry. There is no government intervention in coal prices.

Coal Production: The Coal Memorandum states that the country has considerable reserves but conventional mining methods cannot be used because of high extraction costs. New technologies have to be found but there are no short-term prospects. In order to preserve options for the future an updating of geological knowledge of national coal reserves has been commissioned from the State Geological Office; it will take place in the period up to 1985.

Transportation: The present infrastructure for coal handling and inland transportation is excellent due to very good ports and highly effective waterway facilities. Projects are already taking shape to increase coal import handling facilities: a new coal terminal (Rotterdam) is to be completed by 1983, with a yearly capacity of 8 million tons by 1985 and an option to expand further to 25 million tons/year in the 1990s, depending on the development of world trade.

Facilities in Amsterdam are being expanded from 4 to 6 million tons/year, with improvements of the loading facilities in barges and coasters. A new deep water dry bulkgoods (ore and coal) terminal is planned at the seaside of the approach to Amsterdam with a capacity of 5-10 million tons/year. Although the site has been selected the exact plans have yet to be developed and it is estimated that it will not be in operation before 1987 due to permitting procedures. Both projects will facilitate the handling of the expected growth in coal imports needed for indigenous utilization, and also for transit to the rest of Europe. The railway network is adequate and is well-connected internationally. However, lack of space for new transportation developments could become a constraint in the future.

Trade: The government supports the free flow of coal and international capital in world trade. There is a general policy that a maximum of 20% of coal consumption will be bought on the spot market, with the remaining 80% being covered by long-term contracts and participation in foreign mines. Currently 80% of requirements are covered by contracts. The government is stimulating possible further agreements with coal companies in the United States, Canada and Australia and a first participation in a mine in the United States is nearly completed. Electric utilities, an iron and steel factory and the Dutch State mines (State-owned) have founded a company to contribute to security of supply, through participation in overseas coal production.

Environment: In the Coal Memorandum, some measures are proposed to improve environmental protection. Air pollution and waste disposal are the major areas of concern. The development of new techniques is necessary for a substantial increase in the use of coal.

RD&D: To support the coal policy an important RD&D programme is being developed. The policy is directed mainly to the development of new coal-burning and coal-conversion technologies in order to limit the environmental impact of coal use. The National Research and Development Programme on coal has a 750 million Guilders* budget to promote the commercial use of new coal-burning and conversion technologies such as fluidized bed combustion. The Netherlands Energy Development Company was established in 1976 for the introduction and promotion of large-scale applications of new technologies. The Ministry of Economic Affairs may provide soft loans and guarantees to manufacturers to help finance the development of new products if they complement the National Coal Programme.

Summary of Specific Issues

A Parliamentary decision on energy sources to be used for electricity generation is needed (see Part III of the Energy White Paper) and is the prerequisite for the implementation of the Coal Memorandum. Political constraints related to the coal option have to be faced urgently and realistically because of the long lead times required to build power plants. In this context the difficulty to use coal in electricity generation in conventional ways as a consequence of the environmental standards laid down in the Coal Memorandum should be given continuing consideration. Incentives for encouraging conversion from fuel oil to coal in industry are also needed.

* 1 US $ = 2.50 Guilder as at 31 December, 1981.

NETHERLANDS

Key Solid Fuel and Energy Indicators
(Mtoe)

	1973	1979	1980	1990	Average Annual Growth Rates (%) 1973-1980	1980-1990
Production						
Total Energy Production	57.9	74.5	72.7	47.0	3.3	- 4.3
Solid Fuel Production	1.3	0	0	0	n.a.	n.a.
Solid Fuels as % of Total	2.2	0	0	0		
Trade						
Coal Exports	1.5	1.2	1.6	0.5	0.6	10.8
Coal Imports	3.2	4.9	5.6	13.1	8.1	8.8
Requirements						
Total Primary Energy Requirements	61.9	69.6	65.5	70.3	0.8	0.7
Solid Fuel Requirement	3.2	3.6	4.0	12.6	3.3	12.0
Solid Fuels as % of Total	5.2	5.1	6.2	17.9		
Sectoral Consumption of Solid Fuels and Comparison to Oil						
Solid Fuel Use						
Industry [1]	1.8	1.6	1.7	3.5	- 0.4	7.3
Iron and Steel	1.5	1.4	1.5	n.a.	0.3	n.a.
Other	0.1	0.1	0.1	n.a.	- 5.8	n.a.
Residential and Commercial [2]	0.4	0.1	0.1	n.a.	- 19.6	n.a.
Electric Generation [3]	0.9	1.2	1.9	6.9	11.7	13.9
Solid Fuel Share (%)						
Industry [1]	8.5	5.8	8.4	12.7		
Iron and Steel	54.7	56.5	63.1	n.a.		
Other	1.7	1.0	1.2	n.a.		
Residential and Commercial [2]	1.8	0.4	0.4	n.a.		
Electric Generation [3]	7.0	8.6	13.7	51.9		
Oil Share (%)						
Industry [1]	42.4	49.1	38.5	43.6		
Iron and Steel	19.1	19.0	14.6	n.a.		
Other	28.5	12.7	9.7	n.a.		
Residential and Commercial [2]	34.1	16.4	13.4	n.a.		
Electric Generation [3]	12.9	33.8	37.1	3.8		

1. Excludes non-energy use.
2. Includes Public and Agricultural Use
3. Represents fuel input to electric generation

NEW ZEALAND

Coal Use: Electricity generation is predominantly hydro-based with only one 170 MW power station presently using coal. A further coal-fired station (1000 MW) is now being commissioned and coal will be increasingly used there, with the allocation of natural gas to that station being diverted to other users of gas coming on-stream in the mid-1980s. It is expected that up to 2.6 million tonnes of coal per annum will be required at this station by that date. An additional 500 MW of coal-fired capacity is expected to be installed for commissioning in the early 1990s, and a presently oil-fired thermal station is to be converted to coal with an expected consumption of 125,000 tonnes per annum from 1987.

Following the transfer of one cement works' coal supply from South Island bituminous coals to Waikato subbituminous coals, an additional 70,000 tonnes of coal is required each year from the Waikato field. Expansion of the steel industry is forecast to increase annual demand for subbituminous coal from 150,000 tonnes per annum in 1984 to 900,000 tonnes per annum in 1987.

In several markets, coal use competes with Maui natural gas. Besides the specific advantages of gas compared to coal, constraints to further market penetration by Waikato coal also include the unavoidable time lag before mine output can be increased, and the relatively high costs of distribution.

Energy Prices: Most energy prices continue to be controlled by the government and its pricing policy continues to be that indigenous fuels should be priced so as to reflect as closely as possible their cost of production and to encourage their substitution for imported oil. The second Energy Plan contains more detailed pricing policies and projections for non-oil prices in relation to oil as well as to each other. Indigenous coal prices are based on medium-term marginal costs.

Coal Production: New Zealand has substantial deposits of coal in both the North and South Islands, but high distribution costs and inconvenience in the handling and use of coal (especially in comparison with other fuel sources) have led to a historical decline in its use. However, the steady level of demand during the 1970s is projected to grow over the next fifteen years with increasing sales to large consumers (especially those close to mines) and conversion of coal into more convenient forms.

It is expected that the limited existing production capacity resulting from the past decline in coal use will require long lead times to develop new mines. Only

20% of presently known recoverable coal reserves in the North Island are suitable for mining by opencast methods and it is important that these are carefully managed in order to meet specific and temporary demand shortfalls in coal from underground mines.

Incentives in the form of taxation and asset write-off concessions are provided to encourage development of underground mines in preference to opencast mines. Associated with coal development are problems of capital base, technological development and availability of skilled labour.

Production factors are separate in each of the North and South Islands, with demands placed on each island's coal reserves being affected by different constraints. In the North Island, where substantial amounts of coal will be used as a demand-levelling fuel in electricity generation, there is a necessity to build in some flexibility in production and storage so the supply system can adjust quickly to inflows into the hydro lakes. Any increase above the presently expected demand levels for greater electricity generation and/or steel manufacture will have to be met in the short term from South Island coal or sources other than state coal mines. In the South Island, local demand for coal is expected to be stable, but the proposed level of exports from the west coast will require a considerable increase in total production from that area over the next five years.

This pattern of coal development in New Zealand reflects the role of coal as the marginal fuel — a role assumed until recently by imported oil.

Transportation: In order to keep transportation costs minimal a large proportion of the planned coal usage is to be in industries sited on or near production sources. Where transport is necessary, it is often a significant contributory cost component to the delivered price of coal.

Coal exports originate from the relatively isolated west coast region of the South Island. The lack of a suitable transport infrastructure from this area requires the coal to be railed across to an east coast port (Lyttleton). To increase coal exports significantly, an upgrading of this rail facility would be necessary; and by increasing the coal export trade direct export from the west coast may be made more viable. Options presently being evaluated to increase coal exports include the feasibility of establishing a port facility on the west coast and an investigation of the practicability of a seabed pipeline to an offshore moored buoy loading system.

Trade: New Zealand can now release quantities of coking coal on the international market since it has discovered, in addition to the high quality bituminous coal, reserves of subbituminous coal adequate for its indigenous use. In 1981 coal was exported to Japan, Korea and Fiji. Joint ventures in coal mining, with up to 50% of foreign investment, are examined and judged on a case-by-case basis. Present indications are that foreign investment arrangements can now be made more flexible than was previously the case.

Environment: Land reclamation and rehabilitation obligations are part of all coal mining licence approvals. Emission of dust, noise and effluent are monitored and controlled if judged detrimental.

RD&D: It is intended through RD&D efforts to substantially increase the work on coal conversion to liquid fuels, with emphasis placed on the evaluation of New Zealand's sizeable lignite deposits for this purpose. The current funding level of coal related RD&D is $1.8 million in 1980/81 (12% of energy RD&D budget for 1980/81).

Summary of Specific Issues

Even though substantial coal reserves exist in the country, production is constrained by high costs and the availability of capital and skilled workers. The production projection of 3.9 Mtce in 1990 is, as a consequence, somewhat uncertain. Transportation infrastructure is also a major constraint due to the distant location of mines from major energy consumption centres.

Assessment of resource levels should be accelerated in order to take a decision on the possibility to increase coal production.

NEW ZEALAND

Key Solid Fuel and Energy Indicators
(Mtoe)

	1973	1979	1980	1990	Average Annual Growth Rates (%) 1973-1980	1980-1990
Production						
Total Energy Production	5.8	7.6	7.8	14.3	4.4	6.2
Solid Fuel Production	1.7	1.7	1.9	2.9	1.7	4.4
Solid Fuels as % of Total	28.7	22.6	23.9	20.1		
Trade						
Coal Exports	0	0	0.1	0.4	23.8	16.3
Coal Imports	0	0	0	0	n.a.	n.a.
Requirements						
Total Primary Energy Requirements	9.6	11.1	11.2	16.4	2.2	3.9
Solid Fuel Requirement	1.6	1.7	1.7	2.5	0.8	3.8
Solid Fuels as % of Total	17.0	15.3	15.4	15.3		

Sectoral Consumption of Solid Fuels and Comparison to Oil

	1973	1979	1980	1990	1973-1980	1980-1990
Solid Fuel Use						
Industry [1]	0.9	1.1	1.1	1.2	3.4	0.4
Iron and Steel	0.1	0.1	0.1	n.a.	6.1	n.a.
Other	0.8	1.0	0.9	n.a.	1.3	n.a.
Residential and Commercial [2]	0.2	0.3	0.3	0.2	5.3	-2.3
Electric Generation [3]	0.5	0.2	0.2	1.1	- 11.2	18.0
Solid Fuel Share (%)						
Industry [1]	50.2	40.9	43.3	22.7		
Iron and Steel	100.0	26.7	27.6	n.a.		
Other	48.3	43.5	43.0	n.a.		
Residential and Commercial [2]	8.8	19.4	13.8	13.4		
Electric Generation [3]	10.6	3.9	3.9	13.8		
Oil Share (%)						
Industry [1]	20.6	27.6	25.0	19.2		
Iron and Steel	0	2.0	1.7	n.a.		
Other	21.4	31.8	30.7	n.a.		
Residential and Commercial [2]	48.1	20.0	34.0	5.8		
Electric Generation [3]	7.9	0.4	0.2	0.6		

1. Excludes non-energy use.
2. Includes Public and Agricultural Use
3. Represents fuel input to electric generation

NORWAY

Coal Use: In the long-term coal may be used for electricity generation. In the short and medium term, however, it is not competitive with hydropower which, to a great extent, can still be developed at lower cost than thermal power stations. Increasing oil prices have made coal competitive in the industrial sector, especially where there are relatively large process requirements as in the cement, ceramics and pulp and paper industries.

Energy Prices: Domestic prices of oil products are in line with world market prices. The government supports the view that hydroelectricity prices should be based on long-term marginal costs and this objective should be reached by 1985; substantial nominal price increases will be required if this goal is to be achieved.

Coal Production: Production of coal takes place in Svalbard Island and is mainly sold to Norwegian companies. Small amounts are exported to power plants in Northern Germany. Companies from all countries which have signed the Svalbard treaty may lease land for mining activities. The main constraint is represented by the high exploitation cost of Norwegian reserves. Proven and probable reserves have been estimated at 9 and 16 million tonnes respectively. Over the period 1977-1980, Nkr* 3 million and Nkr 40 million were spent by the government for coal exploration and development respectively.

Transportation: A transportation system to handle the projected volume of imports (2 Mtce in 1990) is now being developed. Furthermore, government and private companies are studying the establishment of a common Scandinavian import centre (where large vessels unload coal for redistribution to final consumers) and the creation of a Norwegian coal transhipment port.

Environment: Mining companies are obliged to observe environmental regulations (to control interference with nature in Svalbard and Jan Meyen Islands) issued by Royal Decree of 28 May 1971. Activities which may harm large areas or which may cause extensive pollution of soil, water or air, can be prohibited. On the use side major constraints are represented by regulations on emission control and ash disposal.

Summary of Specific Issues

To increase the share of coal in the industrial sector above the relatively low share projected for 1990 (15%) requires incentives.

---

* 1 US$ = Nkr 5.86 as at 31 December, 1981.

NORWAY

Key Solid Fuel and Energy Indicators
(Mtoe)

	1973	1979	1980	1990	Average Annual Growth Rates (%) 1973-1980	1980-1990
Production						
Total Energy Production	12.8	52.4	61.4	68.5	25.0	1.1
Solid Fuel Production	0.3	0.5	0.6	0.6	10.3	0.4
Solid Fuels as % of Total	2.3	1.0	0.9	0.9		
Trade						
Coal Exports	0.1	0.1	0.1	0	- 0.4	n.a.
Coal Imports	0.5	0.9	0.9	1.0	6.9	1.3
Requirements						
Total Primary Energy Requirements	19.5	24.2	24.1	26.4	3.0	0.9
Solid Fuel Requirement	0.8	1.3	1.4	1.6	9.7	1.0
Solid Fuels as % of Total	3.9	5.5	6.0	6.1		
Sectoral Consumption of Solid Fuels and Comparison to Oil						
Solid Fuel Use						
Industry [1]	0.7	1.0	1.0	1.2	6.5	1.7
Iron and Steel	0.4	0.7	0.8	n.a.	7.9	n.a.
Other	0.2	0.2	0.2	n.a.	0.5	n.a.
Residential and Commercial [2]	0.1	0.3	0.4	0.4	23.2	0.4
Electric Generation [3]	0	0	0	0	n.a.	n.a.
Solid Fuel Share (%)						
Industry [1]	10.1	13.1	13.8	17.9		
Iron and Steel	35.7	47.5	50.8	n.a.		
Other	5.0	4.6	5.3	n.a.		
Residential and Commercial [2]	2.1	6.5	7.4	6.2		
Electric Generation [3]	0.1	0.1	0.1	0		
Oil Share (%)						
Industry [1]	37.6	40.6	40.7	22.4		
Iron and Steel	8.6	2.6	2.4	n.a.		
Other	40.6	44.8	40.7	n.a.		
Residential and Commercial [2]	54.5	38.8	34.1	30.8		
Electric Generation [3]	0.1	0.1	0.1	0		

1. Excludes non-energy use.
2. Includes Public and Agricultural Use
3. Represents fuel input to electric generation

PORTUGAL

Coal Use: Coal is expected to substitute for oil only for big consumers, starting with electricity generation. Some studies regarding the use of coal in industry have been undertaken and the cement industry, in particular, has plans for conversion. The siting of coal-fired facilities may be a problem. Authorization to build a new coal-fired power station (4 x 300 MW) in the industrial centre of Sines has been obtained with a first unit planned to be fully operational by 1985. However, public opposition has been noted with regard to two other possible sites. CHP has been promoted by a new decree-law which authorizes the building of CHP plants and the selling of surpluses to the public utilities at an equitable tariff. The following fuels may be used:

- residues or by-products (bought or produced);
- renewable energy sources;
- waste heat;
- commercial fuels only when the used technology (CHP) leads to better efficiency than that of the public utilities.

Energy Prices: The fuel oil price is fixed by the government and kept below world levels. This policy is being changed and, in about two years, adjustment to world levels is anticipated. Subsidy of fuel oil recently accounted for about 40% of cost but, by the end of 1981, the fuel oil price was increased by 25% reducing the subsidy to 12.5% of cost. Comparable European prices for fuel oil are in some cases as much as 90% higher than indigenous prices. The price of coal for retail sales (10% of total coal consumed) must be approved by the government. For the remainder of the market, coal prices are unregulated.

Coal Production: Reserves of indigenous coal and lignite are small, of poor quality, and seams are relatively thin. Minerals are exploited under a concession system, accessible to both national and foreign companies, granted by the government. Control of foreign investment in coal production activities is the same as for other economic activities. Studies on planning the enhancement of coal exploration are under way. A feasibility study for the utilization of lignite (Rio Major) for gasification or power generation is under consideration. The actions to enhance national coal resources and develop coal use will be considered under a National Coal Plan, announced by the Minister of Industry and Energy.

Transportation: The country must rely largely on imported coal to meet growing requirements. This implies a substantial change in existing structures, with the need for large investment. Port and transportation infrastructure is not yet in place, although port facilities for large coal carriers and corresponding transshipment are being studied for three major sites (Lisboa, Setubal and Sines) in order to make a rational choice of the site where the main reception port is to be built.

Trade: Long-term contracts for steam coal are to be negotiated by EDP, the public electric utility, for the supply of coal for the Sines power plant from 1985 onwards. Medium-term contracts have been signed for metallurgical coal and possible long-term contracts will depend on approval of the steel industry development plan. A particular constraint on importing coal is the small size of most potential coal users in the industrial sector.

Summary of Specific Issues

Development of coal is subject to the removal of several major constraints. Removal of the price subsidy on fuel oil would increase coal's competitiveness, particularly in the industrial sector. Conversion from oil to coal of one electricity plant located near a port could be possible but is hampered by the vicinity of the town. Construction of new coal-fired power plants is constrained by public opposition but the authorization for building the Sines power plant may ease further development. Long-term contracts for the Sines power plant, currently under negotiation, mention possible delivery date around 1985; this should correspond with the commissioning of the first unit of the plant. Other constraints refer to the construction of main coal ports (e.g. Lisboa, Setubal, etc.) which are still at the study phase as is the improvement of inland transportation. The planning and implementation of coal development could also raise difficult investment issues.

PORTUGAL

Key Solid Fuel and Energy Indicators
(Mtoe)

	1973	1979	1980	1990	Average Annual Growth Rates (%) 1973-1980	1980-1990
Production						
Total Energy Production	1.9	2.8	2.0	4.3	0.9	7.8
Solid Fuel Production	0.2	0.1	0.1	1.2	- 4.1	26.3
Solid Fuels as % of Total	8.2	4.2	5.7	27.8		
Trade						
Coal Exports	0	0	0	0	n.a.	n.a.
Coal Imports	0.3	0.3	0.3	3.9	n.a.	28.0
Requirements						
Total Primary Energy Requirements	7.7	11.0	10.7	21.9	4.9	7.4
Solid Fuel Requirement	0.6	0.5	0.5	5.1	- 3.4	27.0
Solid Fuels as % of Total	7.7	4.3	4.3	23.1		
Sectoral Consumption of Solid Fuels and Comparison to Oil						
Solid Fuel Use						
Industry [1]	0.2	0.3	0.3	2.5	2.8	25.5
Iron and Steel	0.1	0.2	0.2	n.a.	8.0	n.a.
Other	0.1	0	0	n.a.	n.a.	n.a.
Residential and Commercial [2]	0	0	0	0.3	n.a.	n.a.
Electric Generation [3]	0.2	0.2	0.2	1.9	0.0	27.7
Solid Fuel Share (%)						
Industry [1]	8.9	8.0	7.5	31.0		
Iron and Steel	70.4	73.2	73.2	n.a.		
Other	4.6	1.4	0.8	n.a.		
Residential and Commercial [2]	3.7	0.6	0.4	8.8		
Electric Generation [3]	10.7	4.3	4.7	26.0		
Oil Share (%)						
Industry [1]	72.4	71.0	71.8	54.2		
Iron and Steel	17.4	11.8	11.9	n.a.		
Other	75.8	76.3	77.6	n.a.		
Residential and Commercial [2]	66.9	61.8	63.4	61.5		
Electric Generation [3]	14.5	27.1	40.8	31.7		

1. Excludes non-energy use.
2. Includes Public and Agricultural Use
3. Represents fuel input to electric generation

SPAIN

Coal Use: No new oil-fired plant will be authorized (except for specific cases in the islands) and Spain intends to prohibit oil use for power generation on the mainland as of 1985. Dual-fired plants will essentially burn only coal while some natural gas might be used in order to balance the load factor of the new gas transportation grid (Barcelona-Bilbao). An accelerated development plan for new coal-fired plants, based on domestic and imported coal, is being implemented together with the development of reception facilities (in Tarrragona, Algeciras and Gijon). According to this plan, 6500 MW will come into operation by the end of the 1980s in addition to the existing 4700 MW. A conversion plan has been elaborated calling for retrofitting of 1700 MW of oil-fired power plants out of 9300 MW installed capacity by 1987. Utilities which reconvert units to coal may apply for soft loans within the so-called "concerted action" (law to support indigenous coal production and utilization).

A plan for conversion of cement factories from oil to coal has been decided by industry with a total investment of Ptas* 15 billion. Government soft loans are available to cover 70% of investments. Yearly savings of 2 Mtoe of oil are expected compared to 1981 consumption levels. Completion of this conversion is planned for the end of first quarter 1982. Substantial development of coal use will require a restructuring of the refining industry: co-ordination of both processes is a difficult issue.

Energy Price: The price of fuel oil is set by the Administration which aims at following international price movements as closely as possible; a surcharge (500 Ptas or about US$ 5 per ton) is levied on fuel oil for electricity generation and cement factories to encourage fuel switching. The price of indigenous coal for thermal stations is regulated and subsidized and kept competitive with fuel oil; for other uses coal prices are unregulated.

Coal Production: Domestic production is of relatively poor quality and raises technical and economic problems. However, the relatively easy opencast mining, especially for lignite, favours an increase of production in the future. An increase of 6.7% per year is anticipated from 1980 to 1985, when production is expected to have doubled from 1973 levels of 10.3 Mtce. The fiscal regime is based on two laws: the law to encourage mining, which provides tax exemptions, and the mining law, which regulates mining activities and eases administrative procedures. Foreign investment in coal mining is not regulated at levels up to 49% and can exceed this level with government authorization. In addition, increased subsidies to improve production of metallurgical coal have been granted (4525 Ptas/tonne or about US$ 45 per tonne). To avoid decreases in

* 1 US$ = 97.7 Ptas as at 31 December 1981.

domestic coal production, the government has signed an agreement with a group of companies currently producing 89% of steam coal and 93% of hard lignite (subbituminous). This agreement covers the four year period 1981-1985 and involves investment of 80 billion Ptas. The mining companies may get official loans of up to 70% and grants of up to 20% of investment as well as special aids for geological surveys.

Transportation: The current infrastructure is sufficient to handle domestic production. Coal transportation is provided by existing railways and a few small new lines might be added in the future. Harbours, transportation and storage facilities for imported coal are being developed on the Atlantic and Mediterranean Sea (Gijon, Algeciras and Tarragona). These three reception facilities are expected to be capable of handling 17 million tons by 1985-87. The government will not hamper initiatives which may be taken in other areas, for instance in the case of the port of Bilbao.

The civil works expansion of the port of Gijon is envisaged to be completed by mid-1982. The terminal will be able to receive bulk carriers of up to 200,000 dwt and the initial unloading capacity is planned to be 5 million tons/year. For Algeciras and Tarragona the administrative approvals have been granted but development work has not yet started. The first port would be able to accomodate bulk carriers of 100,000-120,000 dwt and receive initially 10 million tons/year. The latter already unloads 3 million tonnes/year with vessels of 60,000 dwt and will be enlarged with two piers for ships up to 150,000 dwt. The expected date of completion of these ports is 1985. These reception facilities should also redistribute imported coal to users located along the coast. The state-owned company CARBOEX, recently created to import coal, is managing the development of reception facilities.

Trade: An aggressive policy to promote coal imports has been decided with the creation of a specific company CARBOEX, owned by the National Industry Institute (Ministry of Industry and Energy). This policy is in the course of implementation. A recently created association of electric utilities will also be active in the international coal market.

Environment: Basic legislation on these matters is being elaborated, which is a constitutional mandate. In this way, a project of law to protect the environment is being prepared which will introduce new measures and update existing legislation. Other laws are being elaborated directed to specific matters. At the same time some work is being done for land reclamation and waste management.

RD&D: Testing of coal/oil mixtures in a coal-fired station is underway. A project to generate 100 MW using fluidized bed combustion is in the planning stage. Testing of atmospheric fluidized bed combustion (AFBC) to burn different domestic solid fuels is foreseen and will be applied to the industrial sector. Other projects to improve the use of solid fuels have already been initiated.

Summary of Specific Issues

Effective conversion from oil to coal in power plants needs to be coordinated with restructuring the refining industry, as the conversion process may be hampered by delays in the adaptation of refinery structures. Fuel switching in industry could be accelerated with government financial aids. A timely conclusion of long-term coal supply contracts, together with an early effective start of development work of the Mediterranean reception facilities (Algeciras and Tarragona), is necessary to meet the import target of 14-15 Mtce by 1990.

SPAIN

Key Solid Fuel and Energy Indicators
(Mtoe)

	1973	1979	1980	1990	Average Annual Growth Rates (%) 1973-1980	1980-1990
Production						
Total Energy Production	16.3	24.3	22.6	46.8	4.7	7.6
Solid Fuel Production	7.2	10.4	12.5	14.8	8.2	1.7
Solid Fuels as % of Total	44.0	43.0	55.4	31.5		
Trade						
Coal Exports	0	0	0	0	n.a.	n.a.
Coal Imports	2.5	3.3	3.8	9.5	6.2	9.6
Requirements						
Total Primary Energy Requirements	57.0	74.2	75.1	106.4	4.0	3.5
Solid Fuel Requirement	10.1	12.5	15.3	24.3	6.2	4.7
Solid Fuels as % of Total	17.7	16.9	20.4	22.8		
Sectoral Consumption of Solid Fuels and Comparison to Oil						
Solid Fuel Use						
Industry [1]	5.4	3.5	4.3	8.4	- 2.9	6.9
Iron and Steel	3.8	2.9	3.1	n.a.	- 2.9	n.a.
Other	1.2	0.4	1.0	n.a.	- 2.5	n.a.
Residential and Commercial [2]	0.4	0.6	0.6	0.8	6.4	3.6
Electric Generation [3]	3.0	7.4	8.9	15.0	16.8	5.4
Solid Fuel Share (%)						
Industry [1]	25.6	14.8	18.2	26.0		
Iron and Steel	67.0	65.0	67.1	n.a.		
Other	10.9	3.4	7.7	n.a.		
Residential and Commercial [2]	4.9	5.4	4.9	4.7		
Electric Generation [3]	17.5	27.8	33.2	34.4		
Oil Share (%)						
Industry [1]	56.8	61.6	57.7	38.0		
Iron and Steel	21.8	15.4	13.1	n.a.		
Other	68.0	68.1	62.5	n.a.		
Residential and Commercial [2]	68.8	69.1	69.2	51.3		
Electric Generation [3]	31.9	24.0	34.2	4.5		

1. Excludes non-energy use.
2. Includes Public and Agricultural Use
3. Represents fuel input to electric generation

SWEDEN

Coal Use: Consumption of coal and coke in 1980 amounted to around 2.6 million tonnes, which corresponds to 5% of total energy consumption. Coal consumption consisted mainly of metallurgical grade coal (2.1 million tonnes) for the iron and steel industry. The remaining 0.5 million tonnes was steam coal, of which industrial consumers such as cement and chemical industries used 0.4 million tonnes and the energy sector around 0.1 million tonnes for combined heat and power plants (CHP).

During the 1980s, the annual consumption of metallurgical grade coal is expected to remain fairly constant. Demand for steam coal in industry is expected to rise to 1-2 million tonnes per year by 1990.

In the energy sector, electricity production is almost entirely based on hydro and nuclear power. There are no coal-fired condensing power plants in the Swedish power system. During the 1980s, the major part of new electricity generating capacity will be nuclear together with some additional hydropower.

In the energy policy decision of 1981, Parliament approved an Oil Substitution Programme aiming at the replacement of 9 Mtoe of oil by other energy sources by 1990. As a result of this, oil imports are expected to decline to 14-17 Mtoe or 40% of the total energy consumption by 1990.

The Oil Substitution Programme consists of the following parts:

— coal	3 Mtoe
— indigenous fuels	3 Mtoe
— heatpumps, solar heating, methanol	1 Mtoe
— electricity for heating purposes	1 Mtoe
— Stockholm district heating	1 Mtoe

The programme will be supported, *inter alia,* by a law stipulating that large new boilers must be designed for the use of solid fuels. The programme is also supported by the Oil Substitution Fund, which offers financial support (grants and loans) for oil substitution projects. The Fund has 1,700 million Sw. Cr*. at its disposal for the years 1981-1983.

A main part of the Oil Substitution Programme is the increase in steam coal consumption. The use of steam coal is planned to grow to 4-6 million tonnes per

* *1 US$ = 5.57 Sw.Cr. as at 31 December 1981.

annum by the end of this decade. This is about half of what is calculated to be potential coal consumption in 1990. There are two reasons for the decision not to try to use the maximum quantity of coal. One reason is the priority given to indigenous solid fuels - peat and wood waste should be used wherever this is technically possible and economically reasonable - and the other reason is the environmental concern. The ambition is to introduce coal in a cautious way to make it possible to use the environmentally best techniques at hand or under development.

Steam coal will be used mainly in industry and in hot-water plants for district heating (DH) during the next 10 years. Because of the favourable electricity production situation, there is no need for taking into operation new big electricity production units during the 1980s and thus there will be no new coal-based condensing plants constructed during this decade.

Energy Prices: Pricing and tax policies are expected to play a fundamental role in achieving the goals of the Energy Bill and the Oil Substitution Programme. It is now a generally accepted rule that the price for different forms of energy should as closely as possible reflect the production and distribution cost to the economy. Price controls for oil were removed in 1980 and oil product prices now generally reflect world market prices for crude oil plus refinery costs in Sweden. Electricity prices are, in principle, freely determined by the utilities and controls by the National Price and Cartel Office aims exclusively at protecting consumers against abusive pricing. No degressive tariffs for large consumers are in practice. As regards coal and peat no established market or pricing policies exist at national level and coal prices reflect import price levels; the small coal tax represents only 3-4% of the end use price and coal is exempted from the much higher value added tax (VAT).

Coal Production: Technically recoverable reserves of hard coal are very small, about 30 million tonnes. The reserves are located in the southern parts of the country. Quality is relatively poor and seams are thin. Presently, only a few thousand tonnes of hard coal are produced each year as a by-product in clay mining for the ceramics industry. As a result, steam coal for the Swedish market must be imported. The policy in this area aims at securing supplies through long-term agreements and participation in coal mines. The instruments for the implementation of this policy are, *inter alia,* credit guarantees for investments and grants for coal exploration.

A state guarantee for investment in coal production was introduced in 1975. The guarantee is also obtainable for investments in the production of oil and natural gas, which were its original purpose. This instrument enables government to support investments by Swedish companies in energy projects abroad that contribute to secure energy supplies to Sweden.

Direct state support for coal exploration has been available since 1981. Companies can obtain grants of 50% of the costs when exploring for coal abroad. These grants are financed by a surcharge on oil products. The surcharge is presently 8 Sw. Cr. per cubic metre of oil. The government has, however, in the 1982 budget proposed an increase to 12 Sw. Cr. per cubic metre, which would bring the Fund about 200 million Sw. Cr. every year. At least during the first half of this decade, however, most of the Exploration Fund built up by this surcharge is planned to be used for oil exploration.

Transportation: A study has been carried out on the requirement for a big central harbour for coal imports to Sweden. The study indicates that due to the cautious approach with respect to the introduction of coal in the energy system, there is at present no need for the government to engage in the construction of a central port for coal imports. The possibility to use joint facilities for coal import to the Scandinavian countries is also being studied.

Trade: The main suppliers of hard coal to Sweden are the United States, the USSR, and the Federal Republic of Germany. Since Swedish coal imports will be rather limited, at least during the 1980s, the government has found it essential for coal importers to cooperate on the world market. The government has given its support to the formation of Swedish Coal Consortium Limited, which is owned by, *inter alia,* the State Power Board and the biggest municipalities.

Environment: Environmental aspects of coal use are very important. Parliament has decided that the environmental situation should be improved when coal is introduced. Because of the severe acidification problem, the main object is to reduce sulphur emissions. In 1981, Parliament limited sulphur emissions of coal-fired plants to 1600 tonnes per year. An even more stringent standard has been proposed by the government. Smaller plants are not allowed to burn coal with a sulphur content higher than 0.6-0.8%.

An important role in this coal-environment work is played by the State Power Board. The Board has been commissioned by the government to investigate solutions for all environmental problems arising from the use of coal and is undertaking a comprehensive study called the "Coal-Health-Environment Project". This work was started in November 1979, and results are reported continuously. A final report is expected in 1983.

RD&D: Because of the necessity to reduce emissions to very low levels, research programmes on coal combustion technologies and gasification techniques have been undertaken. Sweden also participates in coal projects sponsored by the IEA. At present, five fluidized bed combustion units are operating or under construction.

Summary of Specific Issues: Sweden has now to implement the coal policy measures included in the Energy Bill adopted by Parliament in Spring 1981. These measures plan to develop a coal programme which would result in the introduction, by the end of the 1980s, of 4-6 million tonnes of steam coal. Implementation of this programme should proceed with regard to existing constraints, increasingly strict environmental standards, upgrading of existing reception facilities before considering any new coal terminal, and economically and politically acceptable diversification of long-term supply agreements.

SWEDEN

Key Solid Fuel and Energy Indicators
(Mtoe)

	1973	1979	1980	1990	Average Annual Growth Rates (%) 1973-1980	1980-1990
Production						
Total Energy Production	17.8	22.6	23.2	35.9	3.8	4.5
Solid Fuel Production	3.2	3.3	3.2	7.1	0.1	8.2
Solid Fuels as % of Total	18.1	14.5	13.9	19.9		
Trade						
Coal Exports	0	0	0.1	0	n.a.	n.a.
Coal Imports	1.7	1.8	1.7	4.1	0.4	9.1
Requirements						
Total Primary Energy Requirements	47.7	51.6	47.7	58.3	0.0	2.0
Solid Fuel Requirement	4.8	5.0	4.9	11.3	0.3	8.8
Solid Fuels as % of Total	10.0	9.8	10.2	19.3		
Sectoral Consumption of Solid Fuels and Comparison to Oil						
Solid Fuel Use						
Industry [1]	4.0	4.5	4.3	5.6	0.9	2.6
Iron and Steel	0.9	1.1	0.9	n.a.	0.3	n.a.
Other	3.1	3.4	3.3	n.a.	0.9	n.a.
Residential and Commercial [2]	0.4	0	0	2.1	n.a.	n.a.
Electric Generation [3]	0.2	0.2	0.3	0.8	8.0	10.5
Solid Fuel Share (%)						
Industry [1]	25.6	31.5	31.8	34.1		
Iron and Steel	37.0	47.9	46.2	n.a.		
Other	27.7	34.0	34.4	n.a.		
Residential and Commercial [2]	2.8	0.1	0.1	16.1		
Electric Generation [3]	0.9	1.1	1.2	2.6		
Oil Share (%)						
Industry [1]	52.0	43.7	42.0	29.5		
Iron and Steel	42.2	31.7	31.0	n.a.		
Other	49.5	39.7	38.5	n.a.		
Residential and Commercial [2]	80.6	71.8	69.2	17.9		
Electric Generation [3]	14.0	14.1	11.7	2.9		

1. Excludes non-energy use.
2. Includes Public and Agricultural Use
3. Represents fuel input to electric generation

SWITZERLAND

Coal Use: Electricity generation capacity is mainly hydro and nuclear, with a small share of oil-fired capacity (only 4.5% of operating capacity). No new oil-fired power stations are planned. A feasibility study for a coal and gas fuelled 250-350 MW power plant is being carried out for the cantonal authorities of Basel (City and Country). Large industrial energy consumers such as cement, paper and chemical industries have replaced oil by coal or are in the course of doing so for reasons of economics and security of supply. The government has no legal basis for promoting the utilization of coal through financial incentives.

Energy Prices: Energy prices are not controlled and follow the evolution of the market. Coal prices are higher than average European cif prices because of transportation costs.

Transportation: Loading facilities at the Rhine port in Basel exceed today's requirements. The railways' capacities could be a bottleneck if much more coal had to be transported at short notice.

RD&D: One Swiss company participates in an atmospheric fluidized bed combustion project which is part of an IEA R&D programme.

Summary of Specific Issues

The possibility of expanding the coal market needs to be more seriously considered. Coal use could be increased, especially in the area connected to the European waterway system (Basel-on-the-Rhine, major industrial centre of the country). Increased use is possible not only in industry but also in district heating and electricity generation. A major move has been the decision of the cement industry, already implemented, to switch back to coal. Increased coal utilization is constrained by economic considerations stemming from, among other things, the landlocked geographic position of the country.

SWITZERLAND

Key Solid Fuel and Energy Indicators
(Mtoe)

	1973	1979	1980	1990	Average Annual Growth Rates (%) 1973-1980	1980-1990
Production						
Total Energy Production	8.6	10.7	11.6	8.7	4.4	- 2.8
Solid Fuel Production	0.2	0.2	0.2	0.4	- 0.6	5.6
Solid Fuels as % of Total	2.8	2.0	2.0	4.6		
Trade						
Coal Exports	0	0	0	0	n.a.	n.a.
Coal Imports	0.2	0.4	0.5	0.4	11.8	- 2.7
Requirements						
Total Primary Energy Requirements	23.9	24.0	25.0	23.3	0.6	- 0.7
Solid Fuel Requirement	0.6	0.4	0.6	0.8	- 0.1	3.5
Solid Fuels as % of Total	2.4	1.9	2.3	3.4		
Sectoral Consumption of Solid Fuels and Comparison to Oil						
Solid Fuel Use						
Industry [1]	0.1	0.1	0.2	0.3	17.63	2.1
Iron and Steel	0	0	0	n.a.	n.a.	n.a.
Other	0.1	0.1	0.2	n.a.	n.a.	n.a.
Residential and Commercial [2]	0.5	0.3	0.3	0.2	- 4.9	- 4.6
Electric Generation [3]	0	0	0	0.3	n.a.	n.a.
Solid Fuel Share (%)						
Industry [1]	1.9	3.6	6.1	7.3		
Iron and Steel	21.3	15.5	12.7	n.a.		
Other	1.6	3.2	6.1	n.a.		
Residential and Commercial [2]	5.3	3.8	3.9	2.5		
Electric Generation [3]	0	0.1	0.1	3.4		
Oil Share (%)						
Industry [1]	74.7	60.3	58.1	46.3		
Iron and Steel	0	0	0	n.a.		
Other	79.2	68.6	66.5	n.a.		
Residential and Commercial [2]	76.9	71.0	69.2	48.7		
Electric Generation [3]	5.0	3.0	0.6	0		

1. Excludes non-energy use.
2. Includes Public and Agricultural Use
3. Represents fuel input to electric generation

TURKEY

Coal Use: Present policy is to use as much lignite as possible, instead of fuel oil, in every sector. As domestic lignite contains much moisture, sulphur and ash, and has a low calorific value, it is mainly used in electricity generation and is indeed the major fuel input in this sector apart from hydro. Lignite will be utilized at the maximum possible rate and no oil-fired power plants are planned. Coal is recognized as a less expensive energy source than oil but the most important constraint for industry is represented by the large investment needed to allow coal preparation and use (installations for storage, transportation). Cement plants, small industries, the residential/commercial sector (to a certain extent) and possibly district heating and cogeneration are suitable for coal use.

Energy Prices: Prices of all oil fuels, gas and electricity are set by the government, reflecting market conditions. Coal prices are subsidized.

Coal Production: The present policy on depletion of coal reserves is to exploit domestic lignite as far as possible in order to satisfy the growing demand and back out oil. Exploration is under way to discover new coal reserves and research is being undertaken to improve exploitation. Exploitation of coal reserves for heating and electricity generation has priority over industrial use. At present there is no foreign investment in the mining industry. Difficulties in financing projects, in supplying equipment and in finding qualified personnel are major constraints on increasing coal production.

Transportation: Present infrastructure is not adequate to handle projected volumes. The transportation and delivery network must be expanded to meet the requirements anticipated in the development programme. The present transportation system relies mainly on trucks, although studies for improving the railway network are under consideration. In this area investment is a real constraint. The country will import coal in the 1980s, according to the projections, but port reception facilities have to be put in place.

Trade: Efforts will be required to secure the coal imports projected to take place.

Environment: Studies for a policy concerning environmental pollution associated with coal use are currently under way. The Ministry of Industry and Natural Resources issued a recent regulation in order to minimize air pollution.

RD&D: Research on fluidized bed combustion for Turkish lignite, as well as on briquetting and desulphurising of the lignite, is being undertaken. Research on coal liquefaction and gasification is being pursued at the laboratory scale.

Summary of Specific Issues

Turkey has a very ambitious coal production and import programme. It will require substantial effort to secure financing and ensure the priority allocation of scarce foreign exchange. A clear and strong policy will also require a stable climate for domestic and foreign investment in order to make the development of coal mining possible, build the infrastructure and secure the long-term import contracts which are needed to achieve the substantial anticipated coal use. The projected coal consumption in electricity generation by both 1985 and 1990 is partially at risk when considering the generating capacity currently under construction or at the planning stage. Sufficient and timely financing of power generation and mining investments together with intensified efforts to provide and train qualified managerial and technical staff are required if delay in the achievement of these projects is to be avoided.

TURKEY

Key Solid Fuel and Energy Indicators
(Mtoe)

	1973	1979	1980	1990	Average Annual Growth Rates (%) 1973-1980	1980-1990
Production						
Total Energy Production	16.0	18.0	18.3	48.0	1.9	10.1
Solid Fuel Production	11.9	12.8	13.3	32.1	1.6	9.2
Solid Fuels as % of Total	74.1	70.8	72.7	66.9		
Trade						
Coal Exports	0	0	0	0	n.a.	n.a.
Coal Imports	0	0	0	9.6	n.a.	n.a.
Requirements						
Total Primary Energy Requirements	24.8	30.1	31.4	75.4	3.5	9.1
Solid Fuel Requirement	11.9	12.7	13.3	41.7	1.6	12.1
Solid Fuels as % of Total	47.9	42.2	42.2	55.3		

Sectoral Consumption of Solid Fuels and Comparison to Oil

	1973	1979	1980	1990	1973-1980	1980-1990
Solid Fuel Use						
Industry [1]	0.9	2.5	2.5	11.5	15.9	16.3
Iron and Steel	0.5	2.0	1.8	n.a.	19.7	n.a.
Other	0.2	0.4	0.6	n.a.	17.6	n.a.
Residential and Commercial [2]	8.4	9.0	9.6	16.2	2.1	5.3
Electric Generation [3]	1.3	1.8	1.8	13.6	4.5	22.7
Solid Fuel Share (%)						
Industry [1]	23.6	39.1	37.2	48.9		
Iron and Steel	71.6	87.2	88.6	n.a.		
Other	8.0	11.2	15.6	n.a.		
Residential and Commercial [2]	70.5	68.0	68.6	71.7		
Electric Generation [3]	40.5	30.5	29.6	58.0		
Oil Share (%)						
Industry [1]	58.0	44.7	46.7	31.3		
Iron and Steel	21.1	7.3	4.8	n.a.		
Other	69.0	63.6	62.6	n.a.		
Residential and Commercial [2]	26.8	27.4	26.9	20.9		
Electric Generation [3]	40.2	28.2	25.5	9.0		

1. Excludes non-energy use.
2. Includes Public and Agricultural Use
3. Represents fuel input to electric generation

THE UNITED KINGDOM

Coal Use: Coal is highly competitive with oil in electricity generation and about 70% of the domestic market for coal is provided by electricity generation. Oil-fired stations are not being used to meet base load demand and oil burn in electricity generation is presently at about the technical and economic minimum, dictated mainly by the need for system security. No main oil-fired power stations have been authorized since 1973. The National Coal Board (NCB) has an understanding with its major customer, the Central Electricity Generating Board (CEGB), under which the CEGB endeavours to take up to 75 million tonnes of coal per year provided the price increases by no more than the rate of inflation. Other CEGB purchases, including imports, are arranged so as not to jeopardize this understanding. Continued access of imported coal to markets where it is competitive could make a positive contribution to long-term objectives for the industry by reinforcing the incentives to greater efficiency in domestic coal production. It would also help consolidate existing demand and create new demand which eventually could be met by domestic coal.

In industry, coal is also competitive with oil but substitution will be very slow until firms choose to replace oil-fired equipment installed during the 1960s and improved coal combustion techniques become available. As a result, the rate of substitution is expected to accelerate in the second half of this decade. The switch to coal in smaller industrial plants is likely to be constrained by space limitations for coal handling and ash disposal. The current economic recession has reduced coal consumption and likely postponed the conversion process in a great number of industries. As an incentive the government has recently introduced a £50 million* (US$ 95 million) scheme to aid the conversion of industrial boilers to coal firing, providing up to 25% of the capital cost. It is hoped that the scheme will promote a 25% increase in coal use in industry over the next few years. The scheme has recently been widened to ease conversion from gas, as well as oil, to coal. If there is to be significant expansion of demand for domestically produced coal during the present decade, it is likely to come from the industrial sector, perhaps supplemented by expanded exports.

Energy Prices: The government's policy is that the prices of all fuels should be set on an economic basis, each industry taking full account of its costs and market conditions. The government does not set oil product prices. Under the terms of the European Coal and Steel Community Treaty, setting coal prices is

---

*1 US$ = 0.528 £ as at 31 December, 1981

the responsibility of the National Coal Board (NCB). The government has no statutory powers to determine gas and electricity prices; price levels for coal, gas and electricity are, however, importantly influenced by financial discipline imposed by the government on public corporations and utilities; this provides an incentive to take full account of costs and market conditions with the purpose of achieving economic pricing.

Coal Production: The coal industry is almost entirely operated by the National Coal Board, which is state-owned. Government is investing nearly £800 million per year in new mines and restructuring of old mines. The NCB also received government subsidies totalling £254 million in the last financial year (£175 million as operating grants and £79 million as social grants). The results of the investment programme in coal production are becoming visible: in 1980 production was some 7% higher than in 1979. This has, however, coincided with a period of reduced demand for coal, necessitating a revision of the financial strategy aiming to phase out operating grants gradually. As a result, the government recently announced an increased level of financial support for the coal mines. In the immediate future financial support for the NCB will be higher than envisaged, but the government remains committed to a healthy, economically viable industry.

Measures have been adopted to increase coal production and to improve transportation capabilities so that they will be compatible with the needs of the coal-using sectors. The current rate of depletion is not limited by reserves, nor by leasing policy since all coal is vested in the Board. A major problem on the way towards economic viability is the high cost of production in some of the older pits where seams are difficult to work, coupled with constraints on the rate at which production can be concentrated on high productivity, up-to-date pits working in better geological conditions. No distinction is made between production capacity for home and export markets.

Transportation: The NCB considers the existing coal transport and reception system to be satisfactory, although some modifications are under consideration. The principal mode of coal transport is by rail; water-borne transport is used when feasible and cost competitive. Road transport is used mainly for small deliveries from coal concentration depots served by rail.

Trade: Although coal consumers have been free to import coal under open general licence since 1970, the domestic market is in general supplied by indigenous production through negotiated arrangements between the NCB and the Central Electricity Generating Board (CEGB) and industry. Producers are free to export coal.

Environment: Environmental controls on pollution and waste resulting from coal use are common to all industrial operations. The CEGB, as the largest coal consumer, has a specific statutory duty for these matters. A new report on the longer term implications of future coal production and use made by the Commission on Energy and Environment concludes, *inter alia,* that most of the potential environmental problems for increased coal production and use can be overcome by the more even application of techniques and procedures fully understood and currently operational; present legislation will likely afford satisfactory protection of air quality up to the end of the century and beyond. The coal industry is required to restore land after opencast mining and agricultural land must be restored as such.

RD&D: Considerable resources are being devoted to improving coal combustion technology in general and fluidized bed combustion techniques in particular. A demonstration programme for atmospheric fluidized bed combustion systems is under way involving the NCB, plant manufacturers and government, and commercial designs of fluidized-bed-fired boilers and furnaces are becoming increasingly available. An experimental pressurized-bed-combustion facility built under IEA auspices with funds from the United Kingdom, the United States and Germany is in operation at Grimethorpe (Yorkshire). The British Gas Corporation has a substantial programme on the production of substitute natural gas from coal which has proved their Slagging Gasifier at the large pilot scale. A demonstration unit of commercial size is under construction. The NCB research and development programme on coal liquefaction is looking to commercial exploitation of their technology towards the end of the century. Design studies for a pilot plant facility are complete subject to participation by private industry. The NCB is currently negotiating with potential partners.

Summary of Specific Issues

The major current issue is on the demand side, strongly affected by declining economic activity. This would justify strong action to help back out oil in the industrial sector and maintain the coal consumption level in electricity generation at least at current levels throughout the 1980s.

On the production side, coal is affected by a number of high-cost operations which severely affect the national coal industry at a time when international coal trade is expanding. The coal industry is continuing to invest in a massive programme of development and modernization with the aim of regaining viability and making indigenous coal production more competitive with imports. Current efforts to rationalize should be pursued and receive adequate support in order to overcome the current cyclical difficulties even though the rationalization programme may raise difficult social policy issues. Regarding coal imports, although the current situation shows that indigenous supply is plentiful, the possibility to turn towards imports in the longer term should not be disregarded if these imports can be realized under economic long-term conditions.

UNITED KINGDOM

Key Solid Fuel and Energy Indicators
(Mtoe)

	1973	1979	1980	1990	Average Annual Growth Rates (%) 1973-1980	1980-1990
Production						
Total Energy Production	112.3	193.5	198.1	226.5	8.5	1.3
Solid Fuel Production	78.5	70.4	74.8	77.0	- 0.7	0.3
Solid Fuels as % of Total	70.0	36.4	37.8	34.0		
Trade						
Coal Exports	2.0	1.8	3.0	3.5	5.7	1.6
Coal Imports	1.1	2.7	4.3	0	20.8	n.a.
Requirements						
Total Primary Energy Requirements	224.0	220.3	201.5	234.5	- 1.5	1.5
Solid Fuel Requirement	79.0	74.7	69.6	73.5	- 1.8	0.5
Solid Fuels as % of Total	35.3	33.9	34.5	31.3		
Sectoral Consumption of Solid Fuels and Comparison to Oil						
Solid Fuel Use						
Industry [1]	14.3	10.7	6.8	15.0	- 10.1	8.3
Iron and Steel	7.9	6.3	3.2	n.a.	- 12.0	n.a.
Other	6.1	4.2	3.4	n.a.	- 8.0	n.a.
Residential and Commercial [2]	15.1	9.4	8.2	4.0	- 8.3	-6.9
Electric Generation [3]	44.3	53.1	53.1	49.5	2.6	-0.7
Solid Fuel Share (%)						
Industry [1]	21.6	19.2	15.3	25.6		
Iron and Steel	54.4	56.0	50.2	n.a.		
Other	17.7	14.4	13.5	n.a.		
Residential and Commercial [2]	28.1	15.9	14.6	7.0		
Electric Generation [3]	61.1	69.0	73.7	62.7		
Oil Share (%)						
Industry [1]	50.7	40.6	36.8	25.6		
Iron and Steel	31.7	22.1	20.5	n.a.		
Other	51.1	41.4	37.0	n.a.		
Residential and Commercial [2]	24.3	22.9	20.1	12.3		
Electric Generation [3]	26.0	16.0	11.4	12.0		

1. Excludes non-energy use.
2. Includes Public and Agricultural Use
3. Represents fuel input to electric generation

THE UNITED STATES

Coal Use: The government is encouraging the increased use of coal in electricity generation and industrial plants mainly through elimination of oil price controls. As a result, usage of oil in the utility sector has declined and usage of coal has increased significantly as a percentage of total fuel use.

In general, the approach of the Administration in reducing oil use is to allow markets to work and to reduce regulatory obstacles that might prevent a utility from undertaking economic investments. New oil-fired plants are no longer being ordered by utilities because of the high cost of oil, and no such orders are expected except in the possible case of peaking units that might be needed in order to maintain reliable service in case of unexpected high growth. Economics also dictate the use of fuels other than oil in dual-fired plants. Many plants have converted to coal, and many others are in the process of volunteering to convert under the provisions of the Fuel Use Act. The federal government is also issuing exemptions from the Fuel Use Act to allow plants that cannot use coal to switch to natural gas. The existing price differential between coal and oil has resulted in a clear competitive advantage for the use of coal in most new and large boiler applications. This will also be the case in competition with natural gas, prices of which will be decontrolled by 1985. Since utilities must plan new plants almost a decade in advance, displacement of both natural gas and oil is and will remain economic. However, financial constraints aggravated by regulatory obstacles and uncertainty about future demand may prevent utilities from building new plants or completing existing ones in many areas. This accumulation of delays and cancellations together with the linkage between economic growth and electricity usage could result in oil and gas use above optimum levels.

Industry is generally more reluctant than utilities to switch from oil to coal-fired burners and their preference often goes to natural gas as this energy source is still under a price control system. The savings from burning coal are somewhat offset by the high costs of environmental controls and waste disposal, and the energy cost for many industries is often only a small fraction of total production costs. These factors reduce industrial motivation to convert.

The major constraints to increased coal use remain the cost of transportation and environmental considerations.

Energy Prices: Total deregulation of oil prices was implemented in February 1981. The Natural Gas Policy Act of 1978 established a new pricing system for natural gas which will gradually decontrol prices for about half of total gas supplies by 1985, though over a longer time period all gas would be decontrolled as "old gas" production declines. There are no price controls on coal either for domestic or export sales.

Coal Production: The US Geological Survey has identified US coal resources of approximately 1731 billion tons. It is estimated that 438 billion tons — the demonstrated reserve base — was mineable using existing technology. Assuming a conservative recovery percentage of 50 percent, an estimated 219 billion tons of coal can be produced from the demonstrated reserve base. Bituminous coal (229 billion tons; 52%) and subbituminous coal (169 billion tons; 38%) constitute the majority of this demonstrated reserve base. The remainder consists of lignite (33 billion tons; 8%) and anthracite (7 billion tons; 2%). The states west of the Mississippi River contain 54% of the reserve base as well as 84% of US reserves of low-sulphur coal (1% or less sulphur by weight). The current Administration ended a ten-year moratorium on leasing of federal land and is committed to accelerate the federal coal leasing programme and to remove regulatory barriers to coal production. Currently, relevant federal agencies, in conjunction with the Vice President's Task Force on Regulatory Relief, are reviewing the Surface Mining Controls and Reclamation Act (1977), the Federal Coal Management (Leasing) Programme, the Resource Conservation and Recovery Act (1976) and the Clean Air Act, as amended in 1977.

Transportation: The National Energy Transportation Study (NETS) co-sponsored by DOE and DOT and published in July, 1980, assessed the adequacy of the present transportation system to handle expected future increases in the delivery of coal from mine to end user.

Based on the most current EIA estimates* of expected 1990 coal production, the conclusions of NETS still hold. In particular, western coal traffic is expected to increase from 97 million tons in 1975 to nearly 600 million tons in 1990, a six-fold increase. Appalachian coal traffic, while increasing more slowly, will almost double by 1990 to approximately 600 million tons.

Most of this coal will be transported by rail. Currently about two-thirds of all coal is carried by rail and the other third is split between barge and truck transport. Federal regulations regarding rail transport of coal have been relaxed with the Staggers Rail Act of 1980; the law greatly expands rail carrier flexibility by making possible contract service, and by setting rate thresholds below which rates are not subject to ICC (Interstate Commerce Commission) review and possible suspension. The issue remains, however, of coal for export being charged at a rate different from coal for domestic consumption.

Water-only movement of coal is estimated to increase from about 65 million tons in 1975 to about 105 million tons in 1990. The January, 1981 Interim Report of the Interagency Coal Export Task Force points out constraints in the transport system. The upper Mississippi River, in particular Lock and Dam 26, is a bottleneck and this lock is being completely rebuilt. In addition, locks and dams on the Ohio, Kanaawha, and Monongahela rivers have design limitations which are expected to delay barge traffic. The report concludes that, "Those constraints which already exist or are developing are either in the process of improvement or being studied by the Corps of Engineers."

* Energy Information Agency. *Annual Report to Congress,* May 1981.

The US Department of Energy is at the present time developing a policy position concerning the potential technical merits of alternative modes of inland transportation of coal. The analysis of the programmes and policies will take into account existing and potential barriers to implementation of new and/or innovative transportation modes.

Different modes of coal transport are more or less economic in different situations. Because bulk movement of coal by waterways was exempt from ICC regulation under the Transportation Act of 1940, barges have offered both long-term contract and rate advantages. The recent establishment of user fees on the inland waterways will, of course, effect the cost of coal transportation. It is not yet clear, however, if this will alter the competitive position of this mode.

Coal slurry pipelines do have significant right-of-way problems. For the last few years legislation has been introduced in Congress to grant federal eminent domain to interstate pipelines carrying coal. Such legislation has, however, not been enacted. The Administration supports increased competition among the various modes of coal transportation (rail, barge, and pipeline), but opposes giving slurry pipelines rights of eminent domain. In addition to institutional constraints, the economics of slurry pipelines over long distances have yet not been tested and environmental problems referring to water supply need to be overcome (1 ton of coal must be mixed with 1 ton of water in a normal coal slurry application). But slurry pipelines should be allowed to compete fully with other modes of long distance commodity shipments for coal, so that the market can enjoy the potential benefits from this technology.

Port expansion for coal handling has been announced by the private sector in anticipation of a larger volume of coal exports. US coal exporting capacity in 1980 was 130.6 million metric tons per annum (mtpa). However, as of October 1981, nearly 82 mtpa of additional capacity was underway which would increase US capabilities to a minimum of 212.6 mtpa by 1985, a 60% increase. Planned additions to capacity not yet under construction are estimated at about 311 mtpa. Early in 1981, the Inter-Agency Coal Export Task Force presented a survey on potential levels of US coal exports and reviewed the possible constraints on exporting coal from the US as well as options to eliminate or mitigate these constraints. More recently (July 1981) the US Secretary of Commerce announced the Aministration's coal export policy, aimed to expedite issuing permits for and construction of port projects. The Administration is proposing legislation to recover the costs of future deepening and maintenance of channels, through authorising dredging fees to be levied on port users for federal services, payable by local port authorities. This would also allow self-financing public bodies to proceed with port expansion and adaptation without specific and separate authorization by Congress. A positive evolution has been recorded for the port of Baltimore which received, in January 1982, agreement from the federal government to deepen its access to 50 feet, which will allow coal carriers of 100,000-110,000 dwt, fully laden. Final agreement on Baltimore dredging depends on signature of formal contractual agreements by federal and local authorities. Those agreements are still under negotiation. The new loading facilities now under construction are dimensioned in such a way that they will be able to service vessels with substantially higher carrying capacity if they are configured differently (flat bottoms, wider) from current bulk carriers.

Trade: The US will substantially expand coal exports, particularly steam coal. The government encourages the development of long-term coal contracts to facilitate coal trade and development of stable relations between US coal producers and coal importers. Coal export contracts are treated on a national basis and shipments from the producer are delivered on a non-discriminatory basis by destination. US coal production and exports are not regulated by the federal government except in an emergency situation. Despite long coal strikes in 1978 and 1981, the United States government has never interrupted coal trade between its trading partners. Moreover, the current Administration has pledged that it will not interrupt coal trade under contractual commitments unless forced to do so by a severe national emergency, and that it would work with the United States trading partners to minimize adverse impacts in any such event. There are no restrictions on foreign investment in coal production, nor in transportation and export infrastructure development.

Environment: Federal agencies primarily involved with utilising coal in an environmentally acceptable way are the Department of Energy (DOE) and the Environmental Protection Agency (EPA), based on the provisions of the Clean Air Act as amended in 1977. The Departments of Interior and Agriculture also have roles to play. In some instances the Clean Air Act has been faulted for increasing the costs and difficulties of producing and using coal, affecting the siting of major coal facilities, and creating planning uncertainties and construction delays due to complex permit requirements and changeable regulations. It is expected that many coal projects could be unnecessarily and adversely affected by the Act if it stands unmodified. In order to deal with its present inefficiencies, efforts are being undertaken to amend the Clean Air Act, which should result in standards based on more cost-beneficial and effective requirements, streamlined permit procedures, and greater flexibility on the part of regulatory agencies and industrial concerns. At the same time steady progress towards cleaner air would be continued. The Administration is committed to work with Congress to ensure that these principles are considered in the amendments to the Clean Air Act.

Together with the review of the Clean Air Act relevant federal agencies, in conjunction with the Vice President's Task Force on Regulatory Relief, are also reviewing the provisions of the Surface Mining Control and Reclamation Act (1977). The intent is to remove burdensome, excessive and counterproductive requirements.

RD&D: DOE programmes for combustion and heat systems involve exploratory development and technology support of high risk/high pay off approaches to providing a technical foundation adequate to support the increased use of coal. Responsibility for near-term RD&D and the commercial introduction of new technology is left to the private sector.

As regards coal gasification, the government, under the Non-Nuclear R&D Act, has granted a $2.02 billion loan guarantee to the Great Plains Coal Gasification Plant. The Administration has restructured the national synthetic fuels programme to rely more heavily on private investment initiatives. Responsibility for commercializing the technologies of alternative fuels is shifting to the private sector, with potential support from the Synthetic Fuels Corporation. Government funding is ended for demonstration projects which do not muster significant private sector support.

Summary of Specific Issues

Coal production and use constitute a centrepiece of the energy programme in the United States.

Coal production in the United States is constrained only by demand. Lower projected electricity growth rates act to deter further coal penetration in electricity generation. This is compounded by the inability of most power companies to integrate in their rate base the investment in new construction until a plant is effectively producing electricity. Financial constraints and overall financial health of the utilities are under the responsibility of State regulatory commissions. By modifying the rate of return actually earned by utilities which require relief, these commissions can return the utilities to a financial position adequate to undertake economic oil and gas displacement by conversion to coal or by the construction of new coal or nuclear facilities. The move towards full decontrol of all energy prices by the Administration should also act positively to promote coal, as should the Economic Recovery Program designed to reduce taxes to encourage capital investment.

The amendment of the Clean Air Act should provide an opportunity to encourage investment in new coal-fired power plants and conversion of existing facilities in ways compatible with acceptable environmental standards, rather than adding incremental obstacles to these developments.

On the export side, 1980 developments show that great efforts have been made to make maximum use of the loading potential of ports. Though the United States did experience a bottleneck at its leading East Coast coal ports in 1980, private industry responded quickly with an estimated $1 billion investment programme to expand port capacity over the next five years. Moreover, the railroads have developed an advanced ship reservation system whereby vessels can preregister for their place in line and loading dates. These successful results were achieved with many difficulties: congestion of East Coast ports, high demurrage costs, and high freight rates due to the small size of carriers able to call at present port installations. The inland transportation system was also saturated, mainly because of the absence of adequate ground storage facilities at ports.

The substantial export demand currently faced by the United States is likely to be expanded beyond the 1990 anticipated level as a consequence of the substantial increase in import projections of other IEA countries, which can only partially be met by non-IEA supply sources. Future United States coal exports have been substantially increased in recent forecasts and are now in the range of 100-135 Mtce by 1990.

The difficulties currently faced are being considered by both industry and the Administration and approaches to eliminating them need to be actively pursued. An area of specific Administration responsibility is the major issue of dredging access to ports and legislation is currently being proposed to deal with this.

UNITED STATES

Key Solid Fuel and Energy Indicators
(Mtoe)

	1973	1979	1980	1990	Average Annual Growth Rates (%) 1973-1980	1980-1990
Production						
Total Energy Production	1473.2	1522.1	1547.0	1830.0	0.1	1.7
Solid Fuel Production	380.5	440.8	471.1	649.0	3.1	3.3
Solid Fuels as % of Total	25.8	29.0	30.5	35.5		
Trade						
Coal Exports	32.3	39.2	55.1	84.0	7.9	4.3
Coal Imports	0.7	3.6	1.1	n.a.	5.9	n.a.
Requirements						
Total Primary Energy Requirements	1755.3	1874.3	1809.5	2038.0	0.4	1.2
Solid Fuel Requirements	360.0	385.4	405.0	565.0	1.7	3.4
Solid Fuels as % of Total	20.5	20.6	22.4	27.7		
Sectoral Consumption of Solid Fuels and Comparison to Oil						
Solid Fuel Use						
Industry [1]	92.3	80.3	67.2	123.0	- 4.4	6.2
Iron and Steel	46.4	39.7	32.3	n.a.	- 5.0	n.a.
Other	45.8	40.6	34.9	n.a.	- 3.8	n.a.
Residential and Commercial [2]	5.4	4.8	3.7	7.0	- 5.4	6.6
Electric Generation [3]	253.0	294.5	317.9	399.0	3.3	2.3
Solid Fuel Share (%)						
Industry [1]	21.7	20.1	17.7	22.7		
Iron and Steel	59.4	61.3	57.6	n.a.		
Other	15.8	15.9	14.3	n.a.		
Residential and Commercial [2]	1.3	1.1	0.9	1.6		
Electric Generation [3]	49.0	49.7	52.6	50.1		
Oil Share (%)						
Industry [1]	19.4	26.3	26.4	23.4		
Iron and Steel	7.4	8.2	7.7	n.a.		
Other	11.2	14.4	14.0	n.a.		
Residential and Commercial [2]	37.0	30.7	27.1	25.8		
Electric Generation [3]	16.7	13.4	10.4	5.4		

1. Excludes non-energy use.
2. Includes Public and Agricultural Use
3. Represents fuel input to electric generation

B. NON-IEA MAJOR EXPORTING COUNTRIES

REPUBLIC OF SOUTH AFRICA

Overview

The Republic of South Africa is the only industrialized market economy basically fueled with coal. For an array of economic and political reasons, availability of substantial coal reserves, adequate and low-cost labour force and, above all, a determined long-standing attitude of avoiding reliance on imported oil — coal never lost the role of first energy carrier as it did in many other industrialized countries. In 1980 coal supplied almost 75% of total primary energy requirements (TPE) with oil taking 25%. No natural gas or nuclear electricity is currently produced but one nuclear power station (1.8 GW) is under construction.

Total primary energy requirements (TPE) amounted to about 69 Mtoe in 1980, the same order of magnitude as Spain or the Netherlands among IEA Member countries. On a per-capita basis, energy consumption in that same year was 2.82 toe which is relatively low compared to most IEA countries. Growth in TPE has been relatively high during the last decade (5.8% per annum) and did not fall significantly after 1973.

Coal Production

Coal production is presently carried out in the Transvaal, Orange Free State and Natal coal fields, the first being the most important producing region.

The following table provides the evolution of production during the past forty years and anticipated production by the end of the century.

The rapid expansion of production in South Africa started earlier than in most other industrialized countries. The bulk of coal production increase is foreseen over an eight-year period, between 1977 and 1985 with an average anticipated yearly growth of 8.2%. This almost doubling over a short span of time might be compared with several other countries' objectives of doubling production over a ten-year period, but in contrast with many of these projects, most of the increase in South African output is to come from mines already in their pre-production stage.

South African Coal Production

Year	Output (mm. metric tons)	Period	Average Annual Growth in Output (%)
1940	17.2		
1960	38.1	1940-60	4.1
1965	47.6	1960-65	4.6
1970	53.1	1965-70	2.2
1975	69.1	1970-75	5.4
1977	85.4	1975-77	11.2
1979	103.0	1975-79	9.9
1980	113.0	1975-80	10.3
1985	160.0	1980-85	7.2
1990*	200-240	1985-90	4.6-8.5
2000*	260-300	1990-2000	2.3-2.7

Source: Long-term trends in South African coal mining by P. King, *Glückauf* - 116 (1980) Nr. 16.
Statement by Mr. F.W. De Klerk, South African Minister of Mineral and Energy Affairs, 27 May 1981.

* The higher figure takes account of the revised export figure (4 September, 1981).

South Africa's reserve base is substantial and about 51,000 million tons are extractable with current technologies. Most of these reserves contain steam coal and the amount of metallurgical coal reserves is relatively low (700 million tons). This is considered as inadequate in view of the country's very large reserves of ferrous mineral ores which will require processing in the future. The government is at present considering ways and means to safeguard South Africa's scarce availability of coal with strong coking properties.

The seams are generally thick, shallow and nearly horizontal in relatively undisturbed strata. About 96% of the reserves are under less than 200 metres overburden. About 80% of the seams are between 2 and 6 metres in thickness. Currently some 15% of coal extraction is from opencast exploitation, the rest being underground even for some relatively shallow seams which might be worked as opencast mines in most other countries under the same geological conditions. Furthermore, a great share of the underground exploitation is done by the room-and-pillar method, with which recovery of in situ coal may be as low as 10% in extreme cases. In comparison, strip mine recovery rates are over 80%.

The relatively low average levels of coal recovery are mainly due to the following factors:

— South African coal mines tend to be highly labour-intensive. Especially in the past, this was possible with the availability of low-cost labour and also the low level of capital equipment. That in turn is mainly due to coal price controls on the domestic market and the relatively low bargaining power of miners' unions. The trend towards more capital-intensive mining methods is nevertheless evident in the past few years and coincides with the steep increase in coal exports which are priced according to world market level conditions.

— The legislation regarding subsidence prevents the extraction of pillars left, preventing the collapse of the mine roof. With pillar extraction allowed, the recovery rate could be as high as 50-60%.

Bearing in mind the strippable resources, and that opencast mining represents the quickest way to meet expanding demand for coal with a high productivity level and extraction ratio, it is likely that within the next 4-5 years extraction will be raised to 35 million tons/year compared with 24 million tons in 1979. The expansion of opencast mining may be combined with underground exploitation in so-called dual mines.

The quality of South African coal is satisfactory: most of it is classified as steam coal with varying ash content (from 7% for the best anthracites to nearly 30% in some Orange Free State coals). Typical heat content of washed high-grade steam coal that is exported is between 6300-6600 kcal/kg. Sulphur content of steam coal is generally low with a maximum of 1% in weight.

Coal Demand

Roughly 75% of all domestic coal production is for transformation.

(a) Electricity generation

In 1980, 51% of total inland coal consumption went to the power generation sector (43 million tons). In fact, the electric power system is essentially coal-fired. South Africa does not possess adequate hydroelectric resources and no sizeable quantities of oil products are used for electricity generation. Currently there is no nuclear power station in operation but a first plant (Koeberg) with two reactors (2 x 922 MW) is due to come into operation in 1982 and 1983 respectively. The country possesses important uranium reserves and the necessary plants to develop all phases of the nuclear fuel cycle.

Almost 90% of the country's total electricity needs are supplied by a public corporation ESCOM (Electricity Supply Commission) which also owns captive coal mines. The coal is generally burned unwashed in minemouth power plants. The typical heat value for that kind of steam coal is around 5400 kcal/kg.

During the 1970s, generation of electricity increased at an average rate of more than 8% per year. In 1979, total installed capacity was about 16,000 MW including a 540 MW hydropower station and some small gas turbines. The public corporation ESCOM plans to put into operation about 15,000 MW of coal-fired capacity by 1990, doubling the current demand for coal to about 90 million tons per year.

(b) Coke and charcoal production

Production of coke for the iron and steel and ferro-alloy industries was responsible for about 6.4 million tons consumption in 1980. South African coals are not, in general, of very high coking quality and some blending coal has been traditionally imported.

Charcoal is, like coke, a product of coal devolatilization and is used in the manufacture of ferro-alloys and other electric smelting industries. In 1980 more than 0.6 million tons of coal were used in the production of char.

(c) Coal conversion into liquid fuels and gas

South Africa is currently the only country in the world to operate a large scale commercial conversion capacity. The South African Coal, Oil and Gas Corporation (SASOL), founded in 1950, is probably the leading world company in coal gasification and liquefaction. The plants produce a complete spectrum of gaseous and oil products similar to those obtained in a conventional refinery, plus some others produced in petrochemical plants. Feedstocks for the manufacture of nitrogenous fertilizer can also be obtained. Large amounts of gases of different qualities are also produced in the process and are marketed by a SASOL subsidiary. Some of these are distributed as fuel-gas or high-calorie gas to industrial concerns, although the largest share is really a mixture of different gases obtained in the conversion process and marketed as medium calorie gas or coal-gas.

Currently two plants are running, SASOL I and SASOL II. The latter started production in 1980 and in that year 10 million tons of coal went into conversion. SASOL III is under construction and its first phase has started production in the beginning of 1982, and will reach full production in 1983-84. When all three plants are in full operation, a total of 32 million tons of coal will be used for the production of petroleum-type fuels.

(d) Industry and Commerce

Coal use has increased by more than 3.5% per year during the last decade and consumption in 1980 reached about 12 million tons. The industrial sector considered here includes the very important mining sector: in 1980 about 1.5 million tons were used by the mining industry, with gold mining being the most important coal consumer. There are plans to use increasing quantities of coal as raw material for the production of a large range of chemicals. The switch from oil to coal in industry is largely realized and the increased demand for coal in the future will reflect more or less the rate of growth of industry.

Globally the various anticipated demand forecasts of the different sectors between 1980 and 1990 give the following picture:

Evolution of Domestic Demand
(million metric tons)

	1980	1990
Electricity	43.0	85
Coke and Char	7.5	9
Coal conversion	10.0	32
Industry	13.0	19
Miscellaneous	11.0	13
	84.5	158

Source: Long-term trends in South African Coal Mining by P. King, *Glückauf* - 116 (1980) Nr. 16. Industry projection based on 4% growth per year.

Trade

Up to 1973 total exports were around 1.5 million tons, of which more than half was anthracite. However, since 1973, the South African coal export scene has been changing rapidly as is shown below:

Year	Exports (million metric tons)
1973	1.3
1975	2.7
1976	6.0
1977	12.7
1978	15.4
1979	23.4
1980	28.5

In 1976, South Africa was only the seventh largest coal exporter and climbed in 1980 to third place immediately after Poland and the United States. Export quantities are firmly planned to expand to about 44 million tons by 1985-86. In September 1981 it was decided to expand coal exports even further and exports will gradually reach a level of 80 million tonnes/year for a period of 30 years.

Export contracts require government authorization, granted to the mining companies when they plan the opening or expansion of a mine. Most coal exports are based on long- and medium-term contracts. Typical pricing conditions would fix a starting FOB Richards Bay price with escalation clauses to absorb cost increases. No oil price indexation clauses have been reported.

To reach this trade figure an increase of production beyond the projected 1990 level will be needed since comparing the anticipated level of supply and indigenous demand suggests that only 48 million tons would be available for export in 1990. Further production increases are possible, and in the statement made relative to export increase, an upward revision of the resource base has also been announced (51,000 million tonnes). In May 1981, before the upward revision of the resource base a statement by the Minister of Energy projected a production level of 260 million tons for 2000, with a further potential for trebling in the following twenty years.

Infrastructure

Most of the export trade passes through the Richards Bay bulk cargo terminal opened in 1976. This port has a current transshipment capacity of about 25 million tons/year and will be expanded to handle 44 million tons in 1985-86. By that period all the coal exports should transit through Richards Bay except some marginal quantities through Durban. The utilization of Maputo (Lourenzo Marques) as an export harbour is both an acute political and technical issue, although currently about 1 million tons is moving through that port, managed by South African technicians.

Richards Bay is one of the most modern bulk harbours in the world, able to service bulk carriers of up to 150,000 dwt fully laden. The port is connected to the Transvaal coal producing area by a single line railway run by the state-owned railway company (South African Railways). When the line was established all the foundations for the doubling of the line were put in place and the actual doubling will be completed by 1985-86.

Pricing

Coal prices on the domestic market are strictly regulated under the 1964 Price Control Act. The government has the power to fix, on a yearly basis, the maximum price allowable for different coal categories in different producing areas. In June 1981, a representative steam coal price for electricity generation was about 10 Rands (US$ 11.50 per metric ton) but the public electricity corporation ESCOM is buying coal on a contractual basis on the open market at prices below this upper ceiling price. ESCOM is able to exert a strong pressure on prices because of its large captive mines covering a sizeable share of its needs. Higher export prices compensate the companies to a certain extent for the lower domestic price levels. Current export trade is mainly handled by mining companies which developed new mines for that purpose and small mine owners, which until now had no access to export markets. The small mines are requesting part of the new export permits which might be allotted beyond the 44 million tons export level. This would make it easier for them to comply with the strong pressure exerted on their sales prices by ESCOM and generate more income for future development which will be more capital intensive than their current operations.

RD&D

Apart from the large coal conversion plants, another main thrust of RD&D is to improve mining techniques appropriate to the geological condition of the local subsoil, for example through adapting the longwall technique to the specific difficulty of roof carrying because of the hardness of the rock, automatic backfilling of room-and-pillar exploitation with flyash from the nearby mine mouth power plants, and increased extraction rate from room-and-pillar exploitation.

POLAND

Overview

Since 1945, Poland has pursued a policy of maximum utilization of domestic energy resources, relying mainly on hard coal and lignite and some limited resources of natural gas. To balance its energy requirements, Poland is obliged to import crude oil and products, mainly from the USSR. These imports remain, however, below 20% of total primary energy requirements, which amounted to about 125 Mtoe in 1980 (about the same order of magnitude as Italy). On a per capita basis, energy consumption in 1979 was 1.25 toe which is very low compared to most IEA countries. Growth in TPE has been continuously high during the 1970s, averaging 4.2% per annum which is substantially above that for the West European countries during the same decade (2.2% per annum). Neither foreign indebtedness nor current economic difficulties will allow this high annual growth in energy requirements to continue. In the future, efforts will aim at reducing oil and natural gas imports to a minimum, and using a maximum of the only domestic energy source available in large quantities, hard coal and lignite.

Coal Production

Hard coal production is mainly concentrated in Upper and Lower Silesia. Upper Silesia is the most important coal-bearing formation currently under exploitation with coals of different qualities, the rank of coal increasing with the depth of seam. Generally, steam coals occur at lower depth than coking coal. The sulphur content of steam coal is generally low (less than 1%) and steam coal for export has an average calorific content of 6300 kcal/kg.

Lower Silesia is the least important area with respect to the availability of coal and is characterized by exceptionally difficult geological and mining conditions. These coals are, however, worked on account of their high quality for coking. Another rich coal-bearing area in the very Eastern part of Poland, the Lublin area, was in the development stage but work has been stopped since mid-1980. The first commercial output was planned for 1982 and production was gradually to be lifted to a yearly output of 25 million tons by the end of the decade. The achievement of these projections is now postponed for several years.

The table below shows the evolution of production during the past thirty-five years and anticipated production to the end of the century.

Polish Coal Production

Year	Output (Million Metric tons)	Period	Average Annual Growth in Output (%)
1946	47		
1960	104	1946-1960	5.8
1965	119	1960-1965	0.9
1970	140	1965-1970	3.3
1975	172	1970-1975	4.2
1977	186	1975-1977	4.0
1979	201	1975-1979	4.0
1980	193	1975-1980	2.4
1985	235	1979-1985	2.6
1990	260	1985-1990	2.0
2000	300	1990-2000	1.4

Sources: Planning Commission - Council of Ministers (1978) Polish Coal Review *(Weglokoks)* 1980 Vol. 18 nos. 1-3/185-187. 1979 Vol. 17 nos. 1-2/173-174.

The expansion of coal production in Poland has always received maximum government support since it was the major option for developing both the country's industry and a highly regarded export commodity earning hard currency. Until 1980, development plans were realized quite successfully but it has now become doubtful if the pre-1980 production level forecasts will be met. A major change has been the introduction of the five-day week in the coal industry, which has resulted in a downward revision of the production target for 1981: instead of the planned 207 million tons, actual hard coal production reached only 163 million tons. In late 1981, the six day week was reintroduced but Sunday work is now banned and anticipated production for 1982 is set at 180 million tons. Problems about wages do not seem to be as important in miners' claims as issues related to improving working and living conditions and vocational training. More than half the coal industry labour force comes from the agricultural sector and a tendency to go back to the family farm after a short period has developed. A rotation of the labour force of 20% per year or more has become common and this is resulting in a higher proportion of less skilled labour. A consequence is the fluctuation in mining productivity which is a growing concern.

The modernization of mining equipment is also a matter of great concern. Only about 50% of production capacity is mechanized and a great number of extraction machines are often idle for long periods because of lack of spare parts. Another consequence of the gloomy economic situation and the difficulties of raising capital in Western countries, because of high indebtedness, has been the stoppage of development work on the new coal field in the Lublin area.

The projections referring to the development of Polish coal production should presently be considered very cautiously, and the output level of 200 million tonnes might well be very difficult to achieve in the coming years. The table on the previous page should only be considered as an illustration of what the resource basis might allow as potential production.

The need to meet the most urgent domestic energy requirements, linked with the difficulty of obtaining additional oil and gas quantities either from the USSR or on the world market and the reduction of domestic coal production, may constrain the level of coal exports from the usual 40 million tons to below 20 million tons for quite some time.

Coal Demand

The structure of energy consumption in Poland differs from most of the industrialized countries. Coal (hard coal and lignite) covered nearly 94% of energy needs in 1960 and about 75% in 1979. There are three major coal consuming sectors: electricity generation; the steel industry; and residential use.

(a) Electricity generation

Electricity generation is almost entirely based on solid fuels (95% in 1978) and is the most important hard-coal consumer, absorbing about one-third (55 million tons) of total indigenous hard coal consumption (155 million tons). In the future the share of lignite in electricity generation will increase, replacing the high-value Upper Silesian hard coal. Originally 15,000 MW of new capacity were projected to come on stream during the 1980s. It was also planned to reach a nuclear capacity of 500 MW by 1990 but hardly any progress has been recorded in implementing this plan. This is also true of the plan to harness the Lower Vistula river, where 1200 MW of hydro capacity was planned together with a complex and capital-intensive programme of water resource utilization.

In the near future, the bulk of new electricity generation should be covered by the intensive development of lignite resources which is the cheapest option for electricity generation. Failure to push through this development could lead to a continuation of brownout problems that have occurred since 1978.

(b) Metallurgical coal

Poland has developed a large steel industry with integrated coking plants which currently absorb about 25 million tons of hard coal per year. This might rise to 45 million tons per year with the planned expansion of crude steel capacity which foresees doubling of output in the 1980s.

(c) Residential

Poland has a very well developed district heating system in most of the large towns (Warsaw, Lodz, Wroclaw) where more than two-thirds of the energy requirements come from central heating stations. The expected needs for the residential sector are estimated at 30 million tons by 1990 which is only a 3 million ton increase over the 1978-79 figures. The main reason for this small increase is the planned renewal of most of the burning equipment which is out of date and very inefficient. This comment is also valid for electricity generation where the average energy input for producing a kwh is above 3100 kcal, compared to an average input of less than 2200 kcal/kwh in the European Economic Community.

Trade

Poland has been on the export market since the end of the war with a modest 5.5 million tons of exports in 1945 which increased as shown below.

Evolution of Polish Coal Exports

Year	Exports (million metric tons)
1945	5.5
1955	24.0
1960	17.5
1970	28.8
1974	40.0
1975	38.3
1976	38.8
1977	39.2
1978	40.0
1979	41.4
1980	31.0

The official preliminary figures for 1981 are much lower, 15 million tons, of which 7 million tons to Western Europe and 8 million tons to Eastern Europe. For the next five years Poland has announced total exports of 20 million tons/year. The reduction will be applied on a pro-rata basis between West European countries (2/3) and CMEA countries, but there are strong doubts that even this much lower export level will be realized.

As regards export prices, the landed price of Polish steam coal over a long period tends to be close to the landed price of US steam coal, which gives Poland a geographical proximity premium for its exports to West European markets. During the different phases of the oil crisis, when abrupt price increases occurred on the oil market, Polish exporters tried to link their steam coal prices to the evolution of oil prices, especially with countries who were basically dependent on Polish coal exports. This was interpreted at that time as a sign to look for more diversification in coal supplies by the countries affected.

Development of port handling capacity has been adequately realized and PANAMAX type vessels of up to 70,000 dwt can be serviced in two ports (Swinouwicze and Gdansk). The railway link with the coal mines is 800 km long and has always been close to saturation, but with export traffic receiving priority. Current trends in reducing coal exports will leave excess capacity, particularly as regards port infrastructure.

IV STATISTICAL ANNEXES

Annex 1

Coal in IEA Energy Balances 1980-1990

This Annex presents the historical 1980 figures and projections of total primary energy requirements for IEA countries and regions in 1985 and 1990.* The global projections are shown in Table A-1 below.

Table A-1 **IEA Total Primary Energy Requirements**

	1980		1985		1990	
	Mtoe	%	Mtoe	%	Mtoe	%
Solid Fuels	767.7	22	889.3	23	1114.9	27
Oil	1670.3	48	1695.4	44	1554.2	36
Gas	712.3	20	758.2	19	819.6	20
Nuclear	128.8	4	263.4	7	380.9	9
Hydro/Others	235.0	6	274.6	7	338.5	8
Total	3514.1	100	3880.9	100	4208.1	100

Table A-2 presents projections of different energy sources used in electricity generation, industry and residential/commercial sector.

Table A-2 **Fuel Inputs in Electricity Generation**

	1980		1985		1990	
	Mtoe	%	Mtoe	%	Mtoe	%
Solid Fuels	509.5	43	549.2	39	688.6	42
Oil	185.2	15	195.1	14	132.6	8
Gas	128.3	11	118.5	9	132.2	8
Nuclear	128.8	11	263.4	19	380.9	24
Hyd/Geo/Others	235.0	20	264.5	19	299.3	18
Total	1186.8	100.0	1390.7	100.0	1633.6	100.0

* Units employed are million tons of oil equivalent (Mtoe). 1 Mtoe is approximately equal to 1.43 million tons of coal equivalent (Mtce).

Table A-2 (continued) **Energy Use in Industry** [1]

	1980		1985		1990	
	Mtoe	%	Mtoe	%	Mtoe	%
Solid Fuels	175.2	18	257.6	22	308.7	25
Oil	415.2	43	409.9	36	386.3	31
Gas	221.0	23	272.8	24	314.4	25
Electricity and Others	163.9	17	206.9	18	252.0	20
Total	975.3	100.0	1147.2	100.0	1261.4	100.0

Energy Use in Residential/Commercial Sector

	1980		1985		1990	
	Mtoe	%	Mtoe	%	Mtoe	%
Solid Fuels	38.8	5	37.7	4	41.7	4
Oil	292.7	36	301.3	34	268.1	29
Gas	282.3	35	292.8	33	310.8	34
Electricity and Others	197.5	24	247.2	28	303.7	33
Total	811.3	100.0	879.0	100.0	924.3	100.0

1. Includes non-energy use

The projections shown in Tables A-1 and A-2 suggest the following trends:

- — use of solid fuels will increase by 45% between 1980 and 1990.
- — oil and gas requirements together in 1990 are slightly below the 1980 level.
- — nuclear energy will about triple its contribution over the same period.

More detailed information on regional and country breakdowns of projected increases in coal use by sector are provided in the following Annexes. A summary of solid fuel balances for each IEA region up to 1990 is shown in Table A-3.

The Secretariat has also developed, on a preliminary basis, a reference scenario for 1990 and 2000. This has been constructed in the form of complete and consistent energy balances on a regional basis. It thus provides a look at one possible future pattern of production, trade, conversion and use. The methodology employed was to begin with sets of national projections for 1990 and adjust and extend them with a view to indicating what might be achievable in reducing oil use, taking into account technological, economic and institutional factors and allowing for the continued development and implementation of energy policies. The purpose of the scenario is not to present a projection of what is likely to happen, but rather to provide a quantitative and internally consistent framework within which issues of energy policy can be addressed and discussed.

Table A-3 **Solid Fuels Balances for IEA Regions**
Mtoe

	IEA North America			IEA - Europe			IEA - Pacific			IEA - TOTAL		
	1980	1985	1990	1980	1985	1990	1980	1985	1990	1980	1985	1990
Production	496.7	556.7	690.9	206.9	219.8	240.2	74.1	115.0	147.7	777.7	890.9	1078.8
Imports	11.4	9.6	10.3	58.0	71.3	117.0	53.2	70.8	98.6	122.6	151.7	225.9
Exports	- 65.0	- 78.0	- 99.7	- 20.8	- 15.5	- 11.5	- 29.3	- 59.8	- 78.5	- 115.1	- 153.3	- 189.7
Stock Changes	- 12.0	—	—	- 8.5	—	—	3.0	—	—	- 17.5	—	—
Total Requirements	431.2	487.7	601.5	235.7	275.6	345.7	100.8	126.1	167.71	767.7	889.3	1114.9
TPE TOTAL*	2037.7	2158.1	2307.1	1016.6	1140.1	1257.2	459.8	582.6	643.7	3514.1	3880.8	4208.0
% Solids in TPE	21.2	22.6	26.1	23.2	24.2	27.5	21.9	21.6	26.1	21.8	22.9	26.5
Solids in Electricity	332.6	358.7	418.3	139.7	149.0	197.1	37.3	41.5	73.2	509.5	548.2	688.6
TOTAL inputs in electricity*	693.3	793.7	914.7	334.1	382.4	459.3	159.3	214.4	259.5	1186.7	1390.6	1633.5
% Solids in Electricity	48 0	45.2	45.7	41.8	40.0	42.9	23.4	19.4	28.2	42.9	39.4	42.2
Solids in TFC	82.1	123.0	147.2	83.0	102.4	119.6	49.4	70.0	83.7	214.6	295.4	350.5
TFC TOTAL*	1446.6	1506.5	1564.1	735.9	826.9	895.4	324.7	396.1	439.0	2507.2	2729.5	2898.4
% Solids in TFC	5.7	8.2	9.4	11.3	12.4	13.4	15.2	17.7	19.1	8.6	10.8	12.1
Solids in Industry**	78.3	117.9	140.1	52.1	71.0	86.2	44.8	68.7	82.4	175.2	257.6	308.7
INDUSTRY TOTAL**	503.4	566.4	625.8	293.3	355.6	392.0	178.5	225.2	243.7	975.2	1147.2	1261.5
% Solids in Industry	15.6	20.8	22.4	17.8	19.9	22.0	25.1	30.5	33.8	18.0	22.5	24.5

* Memorandum Items
** Industry includes non-energy use.

The reference case for 1990 is slightly higher than the country projections where less coal is assumed to be used in industry and for electricity generation; residential use level in the country projections is slightly higher. By the end of the century the reference case shows coal use increasing from 1979 levels by about 150%, to a level of about 1810 Mtoe (about 2500 Mtce).

Table A-4 **Reference Case Scenario, 1990 and 2000**
(Mtoe)

	1979	1990	2000
Total Primary Energy	3612	4236	5100
Non-Oil Energy Consumption	1794	2666	3780
Oil Consumption	1818	1570	1320
of which: **Net Oil Imports** [1]	1206	974	730
Non-Oil Energy Consumption [2]			
Coal	721 (20.0)	1160 (27.4)	1810 (35.4)
Gas	718 (19.9)	855 (20.2)	965 (18.9)
Nuclear	123 (3.4)	336 (7.9)	555 (10.9)
Hydro	232 (6.4)	285 (6.7)	350 (6.9)
Other	— —	30 (.7)	100 (1.9)
Oil Use			
Industry [3]	475 (26.1)	375 (23.8)	420 (31.8)
Residential/Commercial	325 (17.8)	265 (16.9)	135 (10.2)
Transport	732 (40.3)	707 (45.0)	670 (50.8)
Electricity	216 (11.9)	136 (8.7)	50 (3.8)
Other	70 (3.9)	87 (5.5)	45 (3.4)
Coal Use			
Industry [4]	177 (24.6)	368 (31.7)	485 (26.8)
Residential/Commercial	41 (5.7)	27 (2.3)	35 (1.9)
Transport	1 (.1)	—	—
Electricity	469 (65.0)	705 (60.1)	1095 (60.5)
Other	33 (4.6)	60 (5.2)	195 (10.8)
Memorandum Items			
Net oil Imports (Mbd) [1]	24.5	19.8	14.8
Oil Consumption as % TPE	50.3	37.1	25.9
TPE/GDP Ratio (1973 = 100)	91.2	77.7	68.3
Oil/GDP Ratio (1973 = 100)	88.4	55.4	34.0
Electricity Consumption (Mtoe)	358	494	655
Electricity Cons. as % TFC	13.6	16.9	19.7
Share of Oil in Total Sectoral Energy Use (%)			
Industry [3]	45.6	29.5	25.2
Residential/Commercial	38.7	28.2	19.8
Electricity Generation [4]	18.6	8.6	0.2
Share of Coal in Total Sectoral Energy Use (%)			
Industry [3]	17.0	29.0	29.1
Residential/Commercial	4.9	2.9	3.4
Electricity Generation [4]	40.2	44.6	52.1

Source: IEA Secretariat estimates.

1. Includes marine bunkers. Conversion at 49.2 Mtoe/year = 1 Mbd.
2. Numbers in parentheses indicate share of TPE.
3. Includes non-energy use.
4. Measured on basis of fuel inputs.

The reference case provides a useful overall framework to consider the role of electricity and the coal/nuclear choice. The bulk of the increase in coal use to the end of this century will come in electricity generation, even though load growth assumptions are low by historical standards (an increase of 2.9% per year from 1979 to 2000) and an extremely strong nuclear contribution is foreseen (a four and a half fold increase over the next 20 years). Considering the current uncertainty surrounding both of these issues, there may be further scope than is indicated by the reference scenario for expanding coal use in electricity generation.

Annex 2

Solid Fuel Use in Electricity: Country Data and Projections

Table A-5 shows the historical and projected use of solid fuels in individual countries and by region. Tables A-6, A-7, and A-8 provide supplementary information on electricity generation and capacity, respectively, by energy source.

Table A-5 **Solid Fuel Use in Electricity Generation**

	Mtoe			Share of Fuel Input [1]		
	1973	1980	1990	1973	1980	1990
Canada	8.21	14.70	19.30	12 (4)	16 (3)	16 (2)
United States	253.05	317.93	399.00	49 (17)	53 (10)	50 (5)
NORTH AMERICA	261.26	332.63	418.30	44 (15)	48 (10)	46 (5)
Australia	13.85	22.41	36.70	75 (2)	78 (4)	80 (2)
Japan	12.15	14.63	35.4	11 (68)	12 (41)	17 (11)
New Zealand	0.49	0.21	1.11	11 (8)	4 (0.2)	14 (1)
PACIFIC	26.49	37.25	73.21	21 (56)	23 (33)	28 (15)
Austria	0.70	0.77	1.83	10 (12)	8 (15)	14 (6)
Belgium	2.81	4.58	6.60	28 (50)	35 (30)	37 (10)
Denmark	1.66	5.49	6.80	37 (63)	82 (18)	83 (12)
Germany	47.25	53.99	60.00	67 (13)	64 (6)	52 (3)
Greece	1.52	2.05	8.40	39 (46)	49 (36)	83 (4)
Ireland	0.65	0.59	2.88	34 (57)	22 (53)	66 (15)
Italy	1.59	5.10	22.40	5 (61)	12 (55)	32 (34)
Luxembourg	0.29	0.15	0.20	41 (15)	47 (6)	67 (33)
Netherlands	0.86	1.88	6.90	7 (13)	14 (37)	52 (4)
Norwan	0.02	0.01	—	0.2 (0.1)	0.1 (0.2)	—
Portugal	0.25	0.16	1.90	11 (14)	5 (41)	26 (32)
Spain	2.99	8.89	15.01	17 (32)	33 (34)	34 (5)
Sweden	0.16	0.28	0.75	1 (14)	1 (12)	3 (3)
Switzerland	—	0.01	0.3	— (5)	— (1)	3 (—)
Turkey	1.30	1.76	13.64	40 (40)	30 (25)	58 (9)
United Kingdom	44.35	53.14	49.50	61 (26)	74 (11)	63 (12)
EUROPE	106.40	139.65	197.35	39 (25)	42 (20)	43 (11)
IEA TOTAL	394.15	509.53	688.62	40 (23)	43 (16)	42 (8)

1. Numbers in brackets show the share of oil, for reference.

Table A-6 **Electricity Generation Production** (TWh)

	1980						1990					
	Total Solid Fuels	Oil	Gas	Nuclear	Hydro/ Others	Total	Total Solid Fuels	Oil	Gas	Nuclear	Hydro/ Others	Total
Canada	54.0	15.1	11.0	35.9	252.1	368.1	86.0	13.0	6.0	90.0	340.0	535.0
United States	1210.0	256.3	360.7	266.2	287.4	2380.6	1590.0	163.0	265.0	705.0	354.0	3077.0
NORTH AMERICA	1264.0	271.4	371.7	302.1	539.5	2748.7	1676.0	176.0	271.0	795.0	694.0	3612.0
Australia	66.1	1.2	6.3	—	13.7	87.3	123.0	7.0	12.0	—	17.0	198.0
Japan	45.6	262.5	94.8	82.6	94.2	579.7	99.5	160.0	179.0	253.0	125.0	840.0
New Zealand	0.4	—	1.7	—	20.1	22.2	4.8	0.2	2.7	—	26.6	34.3
PACIFIC	112.1	263.7	102.8	82.6	128.0	689.2	289.8	167.2	193.7	253.0	168.6	1072.3
Austria	3.6	5.4	3.8	—	29.1	41.9	9.3	3.0	6.4	—	39.5	58.2
Belgium	16.0	18.4	5.9	12.6	0.8	53.7	21.1	5.8	2.6	35.0	1.6	66.1
Denmark	21.8	5.3	—	—	—	27.1	24.6	3.4	- 1.0	—	—	30.0
Germany	219.7	25.7	61.0	43.7	18.7	368.8	253.0	16.0	49.0	161.0	19.0	498.0
Greece	9.2	8.6	—	—	3.4	21.2	30.0	1.5	—	—	5.0	36.5
Ireland	1.7	6.4	1.6	—	1.2	10.9	11.5	7.3	2.7	—	0.7	22.2
Italy	19.5	104.6	9.2	2.2	50.2	185.7	95.7	101.0	12.0	35.0	57.3	301.0
Luxembourg	—	—	—	—	0.3	0.3	0.3	0.1	0.2	—	0.1	0.7
Netherlands	9.9	24.9	25.8	4.2	—	64.8	29.4	2.1	18.8	4.2	0.8	55.3
Norway	0.1	—	—	—	84.0	84.1	—	—	—	—	106.0	106.0
Portugal	0.7	6.5	—	—	8.1	15.3	7.9	9.9	—	—	13.2	31.0
Spain	32.4	38.5	3.6	5.2	30.8	110.5	56.1	6.7	—	65.1	44.0	171.9
Sweden	0.8	10.6	—	26.6	58.3	96.3	12.5	2.0	—	54.0	65.5	134.0
Switzerland	0.2	0.6	0.3	14.4	33.9	49.4	0.6	—	0.6	18.6	33.6	53.4
Turkey	6.1	5.8	—	—	11.4	23.3	40.7	6.4	—	2.5	32.1	81.7
United Kingdom	207.5	33.2	2.1	37.0	5.1	284.9	215.5	35.5	—	82.5	4.0	337.5
EUROPE	549.2	294.5	113.3	145.9	335.3	1438.2	808.2	200.7	93.3	457.9	423.4	1983.5
TOTAL	1925.3	829.6	582.0	529.9	1002.8	4876.3	2774.0	543.9	558.0	1505.9	1286.0	6667.8

Table A-7 **Electricity Generation Capacity**
(GW)

	1980						1990					
	Total Solid Fuels	Oil	Gas	Nuclear	Hydro/ Others	Total	Total Solid Fuels	Oil	Gas	Nuclear	Hydro/ Others	Total
Canada	15.0	8.3	4.5	5.9	47.9	81.6	19.3	8.3	4.5	15.7	62.8	110.6
United States	232.0	134.0	77.0	49.0	79.0	559.0	338.0	69.7	113.3	134.0	95.0	750.0
NORTH AMERICA	247.0	142.3	81.5	54.9	126.9	640.6	357.3	78.0	81.5	149.7	157.8	860.6
Australia	14.8	1.8[3]	2.7	—	6.2	25.5	29.7	2.1[3]	3.1	—	7.3	42.2
Japan	5.3	59.5	20.3	15.5	28.8	129.4	23.0	50.3	43.0	46.0	46.7	209.0
New Zealand	0.2	0.7	0.8	—	4.0	5.7	1.2	1.0	0.6	—	5.4	8.2
PACIFIC	20.3	62.0	23.8	15.5	39.0	160.6	53.9	53.4	46.7	52.0	59.4	259.4
Austria	0.7	4.1[1]	—	—	8.2	13.0	2.5	4.5[1]	—	—	11.4	18.4
Belgium	—	8.2	—	1.7	0.5	10.4	5.8	1.6	0.7	5.4	0.4	13.9
Denmark	4.4	1.8	0.3	—	—	6.5	5.7	1.8	0.3	—	0.9	8.7
Germany	42.6	14.7	13.6	9.1	7.3	87.3	54.7	9.8	12.0	26.5	7.8	110.8
Greece	2.3	1.4	—	—	1.4	5.1	5.2	1.2	—	—	4.3	10.7
Ireland	0.5	1.7	0.5	—	0.5	3.2	1.5	1.9	0.7	—	0.5	4.6
Italy	4.7	24.4	—	1.4	16.2	46.4	19.7	33.1	—	5.8	23.5	82.1
Luxembourg	0.1	—	0.2	—	—	0.3	0.1	—	0.1	—	—	0.2
Netherlands	1.6	11.1[1]	2.2	0.5	—	15.4	5.0	7.9[1]	1.4	0.5	0.9	15.7
Norway	—	—	—	—	18.5	18.5	—	—	—	—	29.0	29.0
Portugal	—	1.3[2]	—	—	2.3	3.6	1.8	2.5	—	—	4.3	8.6
Spain	5.3	11.1	—	1.1	13.3	30.8	11.8	7.9	—	12.7	21.4	53.8
Sweden	0.1	7.9	—	4.6	15.1	27.7	1.3	6.1	—	9.5	17.2	34.1
Switzerland	—	0.7	—	1.9	11.5	14.1	0.2	0.7	0.1	2.9	11.5	15.4
Turkey	1.4	1.6	—	—	2.1	5.1	7.0	1.4	—	0.6	9.4	18.4
United Kingdom	47.0	16.8	—	7.7	1.4	72.9	49.1	25.6	—	14.8	1.4	90.9
EUROPE	110.7	106.8	16.8	28.0	98.3	360.3	171.4	106.0	15.3	78.7	143.9	515.3
TOTAL	377.7	311.1	110.1	98.4	264.2	1161.5	582.6	237.4	179.8	274.4	361.1	1635.3

1. Dual-fired (oil/gas);
2. Included 0.15 GW of dual-fired equipment using heavy fuel oil and indigenous coal;
3. Dual-fired (coal/oil) 1980 - 0.4 GW; 1990 - 0.6 GW;

Table A-8 **Share of Electricity Generation Capacity**
(%)

	1980					1990				
	Total Solid Fuels	Oil	Gas	Nuclear	Hydro/Geoth	Total Solid Fuels	Oil	Gas	Nuclear	Hydro/Geoth
Canada	18.4	10.2	5.5	7.2	58.7	17.5	7.5	4.1	14.2	56.8
United States	41.5	24.0	11.6	8.7	14.1	45.1	9.3	15.1	17.9	12.6
NORTH AMERICA	38.6	22.2	10.8	8.6	19.8	41.5	9.1	13.7	17.4	18.3
Australia	58.0	7.1	10.6	—	24.3	70.4	5.0	7.3	—	17.3
Japan	4.1	46.0	15.7	12.0	22.2	11.0	24.1	20.6	22.0	22.3
New Zealand	3.5	12.3	14.0	—	70.2	14.6	12.2	7.3	—	65.9
PACIFIC	12.6	38.6	14.8	9.7	24.3	19.0	20.7	18.4	18.5	23.4
Austria	5.4	31.5	—	—	63.1	13.6	24.5	—	—	61.9
Belgium	—	78.9	—	16.4	4.8	41.7	11.5	5.0	38.9	2.9
Denmark	74.6	20.7	4.7	—	—	65.5	20.7	3.5	—	10.3
Germany	48.8	16.8	15.6	10.4	8.4	49.4	8.8	10.8	23.9	7.1
Greece	45.0	27.5	—	—	27.5	48.6	11.2	—	—	40.2
Ireland	15.6	53.2	15.6	—	15.6	32.6	41.3	15.2	—	10.9
Italy	9.5	52.6	—	3.0	34.9	24.0	40.3	—	7.1	28.6
Luxembourg	33.0	—	67.0	—	—	50.0	—	50.0	—	—
Netherlands	10.4	72.1	14.3	3.2	—	31.9	50.3	8.9	3.2	5.7
Norway	-	—	—	—	100.0	—	—	—	—	100
Portugal	-	36.0	—	—	64.0	20.9	29.1	—	—	50.0
Spain	17.2	36.0	—	3.6	43.2	21.9	14.7	—	23.6	39.8
Sweden	0.4	28.5	—	16.6	54.5	3.8	17.9	—	28.0	50.3
Switzerland	—	5.0	—	13.5	81.6	1.3	4.5	0.7	18.8	74.7
Turkey	27.5	31.4	—	—	41.2	38.0	7.6	—	3.3	51.1
United Kingdom	64.5	23.0	—	10.6	1.9	54.0	28.2	—	16.3	1.5
EUROPE	30.6	29.6	4.7	7.8	27.3	33.3	20.5	2.9	15.3	27.9
TOTAL	32.5	26.8	9.5	8.5	22.7	35.6	14.5	11.0	16.8	22.1

Annex 3

Comparative Economics for Power Generation

The following tables present comparative costs for electricity generation. Table A-9 compares busbar costs in U.S. mills per kilowatt hour for five types of generating plants: nuclear, high-sulphur bituminous coal with scrubbers, low-sulphur sub-bituminous coal with scrubbers, and existing oil and gas generators, sited in the United States. Table A-10 compares for Europe, annualized estimated costs, including capital, fuel costs, operation and maintenance costs for new coal-fired electric generating plants and existing oil-fired power plants. Table A-11 compares indicative total cost estimates for *new* oil, coal and nuclear capacity in the United States, Europe and Japan.

Table A-9 **Comparison of Busbar Costs**
New Plants to Come Online 1989-1998 in United States
(early 1980$)

Costs (Mils/Kwh)	Capital	Fuel	O & M	Total Busbar
Nuclear	21.5-30.2	8.2	2.8	32.4-41.2
Coal (High sulphur bituminous w/scrubbers)	14.8-18.7	14.4-23.6	8.7	37.9-48.3
Coal (Low suplhur sub-bituminous w/scrubbers)	14.6-18.5	9.6-32.6	6.9	31.0-56.1
Existing Oil (1990)*	0	75.3-89.3	1.4	76.7-90.7
Existing Gas (1990)**	0	71.3-88.8	1.3	72.6-90.1

1. Midrange price assumption for residual oil in 1990 is $8.53/mmbtu in 1981$; this equals $7.76/mmbtu in 1980$
2. Midrange price assumption for natural gas in 1990 is $8.00/mmbtu in 1981$; this equals $7.28/mmbtu in 1980$.

Source: U.S. Department of Energy, 1981

Table A-10 Comparison of Powerplant Costs for a New European Coal-Fired Powerplant and an Existing Oil-Fired Plant Assuming a Delivered Coal Price of $60 per Short Ton

Capital Costs	Baseload Operation: Existing Oil	Baseload Operation: New Coal	Intermediate Operation: Existing Oil	Intermediate Operation: New Coal
Capital Cost				
Initial Capital Costs ($/kwh)	0	$1,000	0	$1,000
Real Capital Charge Rate		0.1	0	0.1
Annual Capital Costs		$100		$100
Scrubber Costs (including Capacity Penalty)	0	$200	0	$200
Real Capital Charge Rate	0.125	0.125		
Annual Capital Costs		$25		$25
Total Annual Capital Costs	0	$125	0	$125
Capacity Factor	0.70	0.70	0.50	0.50
House/Year of Generation	6,132	6,132	4,380	4,380
Annualized Capital Costs (Mills/kwh)	0	20.38	0	28.54
Fuel Costs				
Average Fuel Costs ($10⁶ Btu) [1]	$6.00	$2.50	$6.00	$2.50
Heat Rate (Btu/kwh)	9,500	10,000	9,900	10,400
Cost per Kwh (mills/kwh)	57.00	25.00	59.40	26.00
Operation & Maintenance Costs (mills/kwh)	0.5	5.0	0.7	6.0
Total Annualized Costs (mills/kwh)	57.50	50.38	60.10	60.54

1. Assumes a world crude oil price of $35/barrel and a delivered coal price of $60 per ton (assuming 12,000 Btu/lb).
Source: Interim Report of the Interagency Coal Export Task Force, U.S. Department of Energy, January 1981.

Table A-11 Indicative Cost Estimates for Electricity Generation by Fuel
(in 1981 US mills/kwh. Capacity Factor: 65%*)

	Oil 2 × 600 MW		Nuclear	Coal with FGD** 2 × 600 MW		
	Low Sulphur	High Sulphur with FGD	PWR 2 × 1100 MW	US	Europe	Japan
Capital Cost	10.8	12.9	24.8	17.1	17.1	17.8
Operating Cost	2.5	4.2	4.2	5.1	5.1	5.1
Fuel Cost	54.6	47.6	10.0	16.0	26.0	26.0
Total Cost	67.9	64.7	39.0	38.2	48.2	48.9
Reference:						
Capital Investment ($/kw)	577	692	1331	920	920	956
(Initial Investment $/kw)	(500)	(600)	(1000)	(760)	(760)	(790)
(Interest during construction $/kw)	(77)	(92)	(331)	(160)	(160)	(166)
Construction Lead time	3 years	3 years	6 years	4 years	4 years	4 years
Fuel Cost	$33/bbl.	$27/bbl.		$40/t	$65/t	$65/t
($ per toe)	(238)	(194)	(40)	(60)	(100)	(100)
Conversion Efficiency	37%	35%	34%	33%	33%	33%
Heat Rate (k cal/kwh)	2300	2450	2500	2600	2600	2600

Source: World Energy Outlook (Chapter 7; Page 425)
* Capacity factor was chosen to be 65%, considering the current average operation of power plants, either oil, coal or nuclear. Technically, capacity factor could be higher in each case.
** Besides SOx, coal produces various pollutants such as NOx, dust, particulates and ashes. The costs of removing these pollutants to meet legal requirements are included in each cost component.

Annex 4

Coal Use in Industry

Table A-12 **Energy Fuel Use in Industry** [1]
(Mtoe; % shares in parentheses)

	1973	1980	1990
BY SECTOR: TOTAL IEA			
Iron and Steel			
Solid Fuel	125.5 (59.2)	102.4 (61.8)	n.a.
Oil	34.6 (16.3)	17.1 (10.3)	n.a.
Other	51.8 (24.5)	46.2 (27.9)	n.a.
Chemical and Petrochemical			
Solid Fuel	3.6 (1.9)	3.7 (1.7)	n.a.
Oil	135.5 (71.7)	148.4 (69.5)	n.a.
Other	50.7 (26.8)	61.3 (28.7)	n.a.
Other Industry			
Solid Fuel	70.6 (13.3)	69.1 (14.0)	n.a.
Oil	168.5 (31.7)	146.6 (29.7)	n.a.
Other	293.0 (55.0)	277.4 (56.3)	n.a.
BY REGION: TOTAL INDUSTRY			
Europe			
Solid Fuel	60.7 (18)	52.1 (18)	86.2 (22)
Oil	181.6 (54)	134.2 (46)	145.9 (37)
Gas	47.9 (14)	55.7 (19)	84.9 (22)
Electricity and Others	44.8 (14)	51.3 (17)	75.0 (19)
North America			
Solid Fuel	98.2 (19)	78.3 (16)	140.1 (23)
Oil	150.2 (29)	186.2 (37)	145.8 (23)
Gas	194.4 (37)	157.8 (31)	206.3 (33)
Electricity and Others	76.3 (15)	81.2 (16)	133.6 (21)
Pacific			
Solid Fuel	41.0 (24)	44.8 (25)	82.4 (33)
Oil	95.6 (57)	94.8 (53)	94.6 (38)
Gas	3.7 (2)	7.5 (4)	23.2 (9)
Electricity and Others	28.5 (17)	31.5 (18)	49.2 (20)

1. Non-energy use included

Table A-13 **Solid Fuel Use in Industry**[1]

	Mtoe			Share of Solid Fuel in Industry [2]		
	1973	1980	1990	1973	1980	1990
Canada	5.81	11.09	17.10	12 (46)	19 (34)	20 (22)
United States	92.34	67.20	123.00	20 (27)	15 (37)	23 (23)
NORTH AMERICA	98.15	78.29	140.10	19 (29)	16 (37)	22 (23)
Australia	7.40	7.22	13.10	40 (37)	33 (31)	36 (17)
Japan	32.71	36.41	68.1	22 (59)	24 (57)	34 (43) [3]
New Zealand	0.89	1.12	1.17	45 (29)	40 (31)	22 (21)
PACIFIC	41.00	44.75	82.37	24 (57)	25 (53)	34 (39)
Austria	1.44	1.56	1.76	16 (51)	20 (41)	18 (33)
Belgium	5.74	4.86	6.60	30 (41)	31 (30)	33 (31)
Denmark	0.26	0.51	0.90	6 (85)	15 (70)	15 (57)
Germany	18.61	17.27	17.00	20 (51)	22 (42)	19 (34)
Greece	0.58	0.51	1.90	16 (67)	11 (70)	23 (56)
Ireland	0.04	0.01	0.50	3 (83)	1 (67)	13 (55)
Italy	4.62	4.69	7.60	9 (61)	10 (47)	14 (32)
Luxembourg	2.09	1.53	2.00	64 (25)	69 (10)	65 (19)
Netherlands	1.78	1.73	3.50	8 (45)	8 (42)	13 (44)
Norway	0.65	1.01	1.20	9 (43)	13 (46)	16 (32)
Portugal	0.21	0.26	2.47	8 (74)	7 (74)	31 (54)
Spain	5.36	4.35	8.44	22 (62)	17 (61)	24 (42)
Sweden	4.04	4.30	5.57	24 (55)	30 (45)	32 (34)
Switzerland	0.08	0.24	0.30	2 (78)	5 (63)	7 (46)
Turkey	0.90	2.53	11.48	21 (62)	35 (50)	43 (40)
United Kingdom	14.26	6.78	15.00	20 (53)	14 (29)	21 (38)
EUROPE	60.66	52.14	86.22	18 (54)	18 (46)	22 (37)
IEA TOTAL	199.81	175.18	308.69	20 (42)	18 (43)	25 (31)

1. Non-energy use included.
2. Bracketed numbers show the share of oil, for reference.
3. Japan forecasts are Secretariat estimates.

Table A-14 **Solid Fuels in Industry**
(Mtoe)

	1980		1990	
	Coking Coal	Other Coal [1]	Coking Coal	Other Coal [1]
Canada	4.65	1.14	5.80	0.60
United States	29.29	37.91	45.00	78.00
NORTH AMERICA	33.94	39.05	50.80	78.60
Australia	5.75	0.08	11.00	0.10
Japan	30.54	5.87	51.00	17.10 [2]
New Zealand	0.14	0.98	0.12	1.05
PACIFIC	36.43	6.93	62.12	18.25
Austria	1.16	0.13	1.30	0.15
Belgium	4.53	0.33	3.49	3.11
Denmark	—	0.51	—	0.90
Germany	10.20	2.90	10.50	5.50
Greece	0.13	—	0.10	—
Ireland	—	0.01	—	0.50
Italy	4.10	0.49	5.00	2.20
Luxembourg	1.00	0.22	1.90	0.10
Netherlands	1.53	0.10	2.85	0.57
Norway	0.56	0.06	0.60	0.20
Portugal	0.02	0.10	0.06	1.60
Spain	3.78	0.57	4.41	4.03
Sweden	1.32	0.23	1.46	0.65
Switzerland	—	0.24	—	0.30
Turkey	0.51	2.02	4.88	6.60
United Kingdom	—	3.83	4.88	6.60
United Kingdom	—	3.83	6.52	8.89
EUROPE	28.84	11.74	43.07	34.89
IEA TOTAL	99.21	57.72	155.99	131.74

1. Other coal includes steam coal, lignite and brown coal.
2. Japan forecasts are Secretariat estimates.

Annex 5

Solid Fuel and Electricity Use in the Residential/Commercial Sector

Table A-15 **Solid Fuel Use in Residential/Commercial Sector**

	Mtoe			Share of Solid Fuel in Res/Comm Sector [1]		
	1973	1980	1990	1973	1980	1990
Canada	0.40	0.13	0.10	1 (49)	— (37)	— (16)
United States	5.43	3.68	7.00	1 (37)	1 (27)	2 (26)
NORTH AMERICA	5.83	3.81	7.10	1 (38)	1 (28)	1 (25)
Australia	0.66	0.74	0.70	9 (54)	8 (40)	5 (32)
Japan	4.91	3.65	0.40	9 (64)	6 (57)	1 (39) [2]
New Zealand	0.20	0.29	0.23	9 (48)	14 (34)	13 (6)
PACIFIC	5.77	4.68	1.33	9 (62)	7 (54)	1 (45)
Austria	1.52	1.61	1.14	25 (50)	24 (41)	11 (51)
Belgium	2.37	1.08	0.50	18 (64)	8 (53)	3 (47)
Denmark	0.13	0.31	0.60	2 (86)	4 (77)	7 (35)
Germany	10.13	5.94	4.00	13 (65)	8 (54)	5 (46)
Greece	0.06	0.05	—	2 (80)	2 (69)	— (60)
Ireland	0.88	1.28	1.66	5 (48)	47 (34)	48 (25)
Italy	1.56	0.81	1.30	5 (71)	2 (57)	3 (33)
Luxembourg	0.03	0.02	—	5 (79)	3 (63)	— (38)
Netherlands	0.36	0.08	—	2 (34)	— (13)	— (10)
Norway	0.09	0.38	0.40	2 (55)	7 (34)	6 (31)
Portugal	0.04	0.01	0.31	4 (67)	11 (63)	9 (61)
Spain	0.37	0.57	0.81	5 (69)	5 (69)	5 (51)
Sweden	0.39	0.01	2.15	3 (81)	— (69)	16 (18)
Switzerland	0.45	0.32	0.20	5 (77)	4 (69)	13 (49)
Turkey	8.35	9.64	16.23	70 (27)	69 (27)	72 (21)
United Kingdom	15.12	8.22	4.00	28 (24)	15 (20)	7 (12)
EUROPE	41.85	30.33	33.30	16 (55)	11 (45)	11 (33)
IEA TOTAL	53.45	38.82	41.73	7 (46)	5 (36)	5 (29)

1. Bracketed numbers show the share of oil, for reference.
2. Japan forecasts are Secretariat estimates.

Table A-16 **Electricity Use in Residential/Commercial Sector**

	Mtoe			Average Annual Growth (%)	
	1973	1980	1990	1973/80	1980/90
IEA TOTAL	148.3	197.5	266.5	4.2	3.0
North America	97.4	126.5	153.1	3.8	1.9
Europe	38.8	51.0	75.1	4.0	4.0
Pacific	12.1	20.0	38.3	7.4	6.7

Share of Electricity in Residential/Commercial Energy Use (%)

	1973	1980	1990
IEA TOTAL	18	24	29
North America	21	27	30
Europe	15	18	24
Pacific	19	30	33

Annex 6

Coal Production

Table A-17 **Solid Fuel Production and Requirement**
(Mtoe)

	Production		Requirement		Prod/Req. (%)	
	1980	1990	1980	1990	1980	1990
Canada	25.57	41.90	26.15	36.50	98	115
United States	471.11	649.00	405.00	565.00	116	115
NORTH AMERICA	496.68	690.90	431.15	601.50	115	115
Australia	58.95	131.60	32.68	53.40	180	246
Japan	13.24	13.20	66.43	111.80	20	12[1]
New Zealand	1.86	2.86	1.73	2.51	108	114
PACIFIC	74.05	147.66	100.84	167.71	73	88
Austria	1.50	1.00	4.38	5.05	34	20
Belgium	4.44	4.20	11.58	15.00	38	28
Denmark	0.26	0.30	6.14	8.20	4	4
Germany	90.32	89.00	83.50	95.00	108	94
Greece	3.20	8.80	3.52	10.40	91	85
Ireland	1.08	1.73	1.89	5.04	57	34
Italy	1.31	2.00	13.24	34.00	10	6
Luxembourg	—	—	1.81	2.20	—	—
Netherlands	—	—	4.05	12.60	—	—
Norway	0.58	0.60	1.44	1.60	40	38
Portugal	0.12	1.19	0.46	5.05	26	24
Spain	12.50	14.76	15.31	24.26	82	61
Sweden	3.23	7.14	4.87	11.27	66	63
Switzerland	0.23	0.40	0.57	0.80	40	50
Turkey	13.31	32.11	13.26	41.70	100	77
United Kingdom	74.84	77.00	69.62	73.50	107	105
EUROPE	206.92	240.23	235.64	345.67	88	70
IEA TOTAL	777.65	1078.79	767.63	1114.88	101	97

1. Japan forecasts are Secretariat estimates.

Table A-18 **Coking Coal Production**
(Mtoe)

			Difference 1980-1990	
	1980	1990	Mtoe	% Increase
IEA Europe	44.3	44.8	0.5	1.1
IEA North America	84.3	104.9	20.6	24.4
IEA Pacific	38.8	55.6	16.8	43.3
IEA Total	167.4	205.3	37.9	22.6

Table A-19 Coking Coal Production and Requirements

	Production		Requirement		Prod/Req (%)	
	1980	1990	1980	1990	1980	1990
Canada	9.33	13.90	4.64	5.80	201.1	239.7
United States	75.00	91.00	37.35	48.00	200.8	189.6
NORTH AMERICA	84.33	104.90	41.99	53.80	200.8	195.0
Australia	32.52	50.50	9.63	11.10	337.7	455.0
Japan	5.98	4.60	54.29	63.60	11.0	7.2[1]
New Zealand	0.32	0.47	0.20	0.12	160.0	391.7
PACIFIC	38.82	55.57	64.12	74.82	60.5	74.3
Austria	—	—	2.24	3.14	—	—
Belgium	2.76	2.60	5.08	5.87	54.3	44.3
Denmark	—	—	—	—	—	—
Germany	32.00	31.00	17.10	24.00	187.1	129.2
Greece	—	—	0.32	0.40	—	—
Ireland	—	—	—	—	—	—
Italy	—	—	7.71	8.60	—	—
Luxembourg	—	—	1.56	1.90	—	—
Netherlands	—	—	2.88	5.65	—	—
Norway	—	—	0.60	0.60	—	—
Portugal	—	—	0.29	1.24	—	—
Spain	1.48	1.96	4.63	4.41	32.0	44.4
Sweden	—	—	1.32	1.46	—	—
Switzerland	—	—	—	—	—	—
Turkey	0.98	1.99	0.98	5.22	100.0	38.1
United Kingdom	7.06	7.26	7.96	7.22	88.7	100.6
EUROPE	44.28	44.81	52.67	69.71	84.1	64.3
TOTAL	167.43	205.28	158.78	198.33	105.4	103.5

1. Japan forecasts are Secretariat estimates.

Table A-20 Other Coal Production
(Mtoe)

			Difference 1980-1990	
	1980	1990	Mtoe	% Increase
IEA Europe	147.6	163.9	16.3	11.0
IEA North America	406.7	570.0	163.3	40.2
IEA Pacific	32.8	89.6	56.8	173.2
IEA Total	587.2	823.5	236.3	40.2

Table A-21 **Other Coal Production and Requirements**

	Production		Requirement		Prod/Req (%)	
	1980	1990	1980	1990	1980	1990
Canada	10.60	17.00	15.87	19.70	66.8	86.3
United States	396.11	553.00	368.06	517.00	107.6	107.0
NORTH AMERICA	406.71	570.00	383.93	536.70	105.9	106.2
Australia	24.26	78.60	20.88	39.80	116.2	197.5
Japan	7.03	8.60	11.91	48.20	59.0	17.8[1]
New Zealand	1.53	2.36	1.51	2.36	101.3	100.0
PACIFIC	32.82	89.56	34.0	90.36	96.5	99.1
Austria	0.68	0.45	1.30	1.32	52.3	34.1
Belgium	1.68	1.60	6.51	9.13	25.8	17.5
Denmark	—	—	5.88	7.90	—	—
Germany	57.60	58.0	65.70	71.00	87.7	81.7
Greece	—	—	—	—	—	—
Ireland	0.04	0.04	0.84	3.35	4.8	1.2
Italy	0.39	0.90	4.62	24.30	8.4	3.7
Luxembourg	—	—	0.25	0.30	—	—
Netherlands	—	—	1.04	6.81	—	—
Norway	0.21	0.30	0.15	0.30	140.0	100.0
Portugal	0.01	0.09	0.07	2.71	14.3	3.3
Spain	11.02	12.80	10.68	19.85	103.2	64.5
Sweden	0.01	—	0.33	2.67	3.0	—
Switzerland	—	—	0.34	0.40	—	—
Turkey	8.22	19.95	8.17	26.31	100.6	75.8
United Kingdom	67.78	69.74	63.26	67.16	107.1	103.8
EUROPE	147.64	163.87	169.14	243.51	87.3	67.3
TOTAL	587.17	823.43	587.07	870.57	100.0	94.6

1. Japan forecasts are Secretariat estimates.

Annex 7

Coal Trade

Table A-22 **Total Coal Trade** (Mtoe)

	Exports		Imports		Net Imports or Exports (-)	
	1980	1990	1980	1990	1980	1990
Canada	9.91	15.70	10.34	10.30	0.43	- 5.40
United States	55.07	84.00	1.09	—	- 53.98	- 84.00
NORTH AMERICA	64.98	99.70	11.43	10.30	- 53.55	- 89.40
Australia	27.87	78.10	0.01	—	- 27.86	- 78.10
Japan	1.39	—	53.15	98.60	51.76	98.60[1]
New Zealand	0.08	0.35	—	—	- 0.08	- 0.35
PACIFIC	29.34	78.45	53.16	98.6	23.82	20.15
Austria	0.02	0.01	2.91	4.06	2.89	4.05
Belgium	0.87	0.50	8.05	11.30	7.18	10.80
Denmark	0.04	—	6.10	7.90	6.06	7.90
Germany	14.55	7.00	8.09	13.00	- 6.46	6.00
Greece	—	—	0.40	1.60	0.40	1.60
Ireland	0.04	—	0.84	3.31	0.80	3.31
Italy	0.52	—	12.62	32.00	12.10	32.00
Luxembourg	—	—	1.82	2.20	1.82	2.20
Netherlands	1.57	0.50	5.62	13.10	4.05	12.60
Norway	0.09	—	0.88	1.00	0.79	1.00
Portugal	—	—	0.33	3.86	0.33	3.86
Spain	—	—	3.81	9.50	3.81	9.50
Sweden	0.08	—	1.72	4.13	1.64	4.13
Switzerland	—	—	0.53	0.40	0.53	0.40
Turkey	—	—	—	9.59	—	9.59
United Kingdom	3.00	3.50	4.33	—	1.33	- 3.50
EUROPE	20.78	11.51	58.05	116.95	37.27	105.44
TOTAL	115.10	189.66	122.64	225.85	7.54	36.19

* Japan forecasts are Secretariat estimates.

Table A-23 **Coking Coal Trade** (Mtoe)

	Exports		Imports		Net Imports or Exports (-)	
	1980	1990	1980	1990	1980	1990
Canada	9.15	13.50	4.16	5.40	- 4.99	- 8.10
United States	38.65	43.00	—	—	- 38.65	- 43.00
NORTH AMERICA	47.80	56.50	4.16	5.40	- 43.64	- 51.10
Australia	23.73	39.40	—	—	23.73	- 39.40
Japan	1.39	—	48.27	59.00	46.88	59.00[1]
New Zealand	0.08	0.35	—	—	- 0.08	- 0.35
PACIFIC	25.20	39.75	48.27	59.00	23.07	19.25
Austria	—	—	2.25	3.14	2.25	3.14
Belgium	0.87	—	3.22	3.27	2.35	3.27
Denmark	—	—	—	—	—	—
Germany	14.55	7.00	—	—	- 14.55	- 7.00
Greece	—	—	0.40	0.40	0.40	0.40
Ireland	—	—	—	—	—	—
Italy	—	—	7.71	8.60	7.71	8.60
Luxembourg	—	—	1.56	1.90	1.56	1.90
Netherlands	—	—	2.88	5.65	2.88	5.65
Norway	—	—	0.52	0.60	0.52	0.60
Portugal	—	—	0.27	1.24	0.27	1.24
Spain	—	—	3.15	2.45	3.15	2.45
Sweden	0.08	—	1.40	1.46	1.32	1.46
Switzerland	—	—	—	—	—	—
Turkey	—	—	—	3.23	—	3.23
United Kingdom	0.04	0.04	1.49	—	1.45	- 0.04
EUROPE	15.54	7.04	24.85	31.94	9.31	24.90
TOTAL	88.54	103.29	77.28	96.34	- 11.26	- 6.95

1. Japan forecasts are Secretariat estimates.

Table A-24 **Other Coal Trade** [1]
(Mtoe)

	Exports		Imports		Net Imports or Exports (-)	
	1980	1990	1980	1990	1980	1990
Canada	0.78	2.20	6.18	4.90	5.40	2.70
United States	15.46	41.00	0.54	—	- 14.92	- 41.00
NORTH AMERICA	16.24	43.20	6.72	4.90	- 9.52	- 38.30
Australia	4.14	38.70	0.01	—	- 4.13	- 38.70
Japan	—	—	4.88	39.60	4.88	39.60[2]
New Zealand	—	—	—	—	—	—
PACIFIC	4.14	38.70	4.89	39.60	0.75	0.90
Austria	0.01	0.01	0.63	0.88	0.62	0.87
Belgium	—	0.50	4.83	8.03	4.83	7.53
Denmark	0.04	—	6.10	7.90	6.06	7.90
Germany	—	—	8.09	13.00	8.09	13.00
Greece	—	—	—	—	—	—
Ireland	0.04	—	0.84	3.31	0.80	3.31
Italy	0.52	—	4.91	23.40	4.39	23.40
Luxembourg	—	—	0.26	0.30	0.26	0.30
Netherlands	1.57	0.50	2.61	7.31	1.04	6.81
Norway	0.06	—	—	—	- 0.06	—
Portugal	—	—	0.06	2.62	0.06	2.62
Spain	—	—	0.66	7.05	0.66	7.05
Sweden	—	—	0.32	2.67	0.32	2.67
Switzerland	—	—	0.53	0.40	0.53	0.40
Turkey	—	—	—	6.36	—	6.36
United Kingdom	2.21	2.58	2.74	—	0.53	2.58
EUROPE	4.45	3.59	32.58	83.23	28.13	79.64
TOTAL	24.83	85.49	44.19	127.73	19.36	42.24

1. Other coal refers mainly to steam coal.
2. Japan forecasts are Secretariat estimates.

Annex 8

Table A-25 Steam Coal Prices [5] as of Late December 1981
(in US$ per ton)

	FOB Price Coast [1]	Maritime Freight [2]	Europe av. CIF Price	Japan av. CIF Price [5]	Europe av. CIF Price in US $ per toe	Japan av. CIF Price in US $ per toe	Ratio coal to Spot HFO (1%S) Rotterdam 210$/t	Ratio coal to Spot HFO (1%S) Japan* 240$/t	Cal val. kcal/kg.	Sulph. cont.%
Australia									6400	1
Europe	50-54	12-17	62-71		97-111		0.46-0.48			
Japan	50-54			70		109		0.41-0.45		
Canada									6600	0.5
Europe	46-48	16-18	62-66		94-100		0.45-0.48			
Japan	46-48			65		98				0.35-0.41
Poland [3]									6300	1
Europe										
Japan										
R. South Africa									6100	0.5-1.2
Europe	40-45	12-14	52-59		85-97			0.40-0.46		
Japan	40-45			58		95				0.35-0.42
United States [4]									6250	1 -2
Europe	48-50	7- 9	55-59		88-94		0.48-0.45			
Japan	48-50			75		120			0.43-0.50	

1. Estimates from various trade sources.
2. Average of single trip ocean freight rates for vessels ranging between 50000 and 100000 dwt.
3. Exports of steam coal almost nil in 1st quarter 1981.
4. Maximum size accessible are Panamex type vessels (65000-75000 dwt)
5. As for Japan CIF coal prices are from Japanese custom data. Other data are from various but not authoritative sources and should be taken just for illustrative purposes.

* Price base of LSF in Singapore plus freight to Japan.

V APPENDICES

Appendix 1

STATEMENT OF POSITION THE COAL INDUSTRY ADVISORY BOARD 29 APRIL 1982

1. The CIAB is greatly concerned about the situation that is developing as the real price of oil continues to drop and oil remains in abundant supply and coal prices have continued to rise. The impression that the CIAB got from the IEA's Conclusions of the Coal Review is that the Coal Review does not sufficiently elaborate upon overall strategic considerations in supporting a programme to increase coal use so as to reduce oil consumption. The present oil and coal price trends are raising serious questions regarding whether or not, in fact, the economics justify, by themselves, an aggresssive programme to displace oil in both electric utilities and industrial companies. It is the view of the CIAB that strategic issues are overwhelmingly important and should be emphasized as basic to the programme. There are danger signals that four factors could reduce the pace of substitution of coal for oil:

— the fluctuating oil prices themselves;
— rising coal prices from some sources;
— the present recession which has reduced dramatically the overall demand for energy;
— the continuation of high interest rates which make investment in new plant and equipment or in conversion equipment expensive to finance.

We are concerned that a return to reliance upon oil, which may not be economically disadvantageous at the present time, will simply result in a repetition of the oil shocks of 1974 and 1979 and may, in addition, have the affect of returning the industrial countries into recession. This latter fear is based on a concern that, with continued reliance on oil, a resurgence of the Western economies will increase oil demand and with a tightening of the oil markets prices will return to the levels of 1981. Then, the higher energy prices will drive the economies back into recession.

Coal production, transportation and use must satisfy acceptable environmental standards. However, there may be a tendency to cede the entire economic gap between coal prices and oil prices to environmental controls. As well, coal price increases which in some cases arise through government fiscal or regulatory action also tend to narrow this gap. When the gap is narrowed, whether through dropping oil prices, rising coal prices or increased environmental control costs, the economic incentive to replace oil with coal is lessened.

2. The CIAB is concerned that the presence of IEA forecasts of increased coal use as presently contained in the Conclusions document create the impression that the coal to oil substitution battle has been won and the path is well defined. The forecasts look very encouraging but the critical question is whether or not they will be realized. The CIAB believes that further efforts will be needed and these efforts will require, most importantly, the development of stable long-term relationships between buyers and sellers, through long-term contracts, joint ventures or other appropriate means. Producers will have to continue to make efforts to minimize costs and provide stable supplies at competitive prices; users must clarify the role they expect coal to play in their energy mix and take necessary steps, including investment, to bring this about. We believe that the IEA should continue to review the situation in individual countries on a regular basis in order to focus attention on whether the forecasts are likely to be realized without additional policy action.

3. The third issue related to priorities regarding research, development and demonstration. While the CIAB recognizes that IEA Member countries vary widely with respect to whether or not it is proper for governments to encourage research in new technologies that the markets are not encouraging, we also believe that strategic imperatives are not recognized by the market. The CIAB believes that more consideration needs to be given by the IEA Member countries, acting separately and in cooperation, to the level of commitment and direction of strategically oriented RD&D and to how such RD&D should be sponsored and funded.

4. The fourth, and last, issue the CIAB wishes to highlight is a general public perception of the basic credibility and mission of the IEA itself. The IEA was basically created as an instrument for reducing oil dependency among the Member countries. We recognize that, given the present situation in the oil markets, the IEA has a difficult task to demand attention in government and industry circles. However, for the reasons stated with respect to the first issue recommended for discussion, we believe that the IEA's task, and by inference, the CIAB's task, may now be more important than it ever was in the past.

Given the present circumstances and the overall conclusions of the Coal Review, the CIAB suggests that the IEA give greater focus in its annual reviews and continuing analysis of the coal situation to the development of coal policies in accordance with the Principles for IEA Action on Coal. In particular, it would be desirable to focus on the performance by:

1. **Users,** to continue their resolve to reduce dependence on oil despite short-term uncertainties.

2. **Suppliers,** to concentrate on minimizing costs in an effort to hold margins against a fluctuating oil price.

3. **Governments,** to take heed of the dangers of eroding coal's market advantage through taxes, government charges, environmental restrictions and planning delays.

4. **All,** to seek ways of monitoring the funding of long-term R&D which has strategic and political implications over and above market conditions.

Appendix 2

PRINCIPLES FOR IEA ACTION ON COAL*

The Governing Board at Ministerial Level

considering:

The Agreement on an International Energy Program, signed in Paris on 18th November, 1974 and the Participating Countries' determination to reduce over the longer term their dependence on imported oil for meeting their energy requirements and to encourage the development of alternative sources of energy such as coal, as provided in Articles 41 and 42 of that Agreement;

The Long-Term Co-operation Programme of the International Energy Agency, adopted by the Governing Board on 30th January, 1976, which provides for the accelerated development of alternative sources of energy; for co-operation in specific energy sectors, including coal; and for other long-term energy measures;

The Ministerial Decision on Group Objectives and Principles for Energy Policy, adopted by the Governing Board meeting at Ministerial Level on 5th and 6th October, 1977; and that it is now necessary to elaborate in more detail the action called for by Principle No. 6 of those Principles for Energy Policy, providing for a strong steam coal utilization strategy and active promotion of an expanded and reliable international trade in steam coal;

The Recommendation on Coal and the Environment adopted by the OECD Council on 8th May, 1979 following a recommendation by the Environment Committee meeting at Ministerial Level on 7th-8th May, 1978.

I. Coal in the Overall Energy Context

The Governing Board at Ministerial Level:

1. Has reassessed the overall energy prospects to 1985 and beyond, and concludes:

    a. The basic prospects for energy market developments have not improved since the meeting of the Governing Board at Ministerial Level in October, 1977.

---

* Governing Board meeting of 21-22 May 1979.

b. The world is still confronted with the serious risk that within the decade of the 1980's it will not have sufficient oil and other forms of energy available at reasonable prices unless present energy policies are strengthened. Such a situation would have severe economic, social and political repercussions in all Member countries and throughout the world.

c. Developments in international oil markets during recent months have demonstrated how difficult it is to predict accurately when such a situation might arise. They have, moreover, shown clearly how exposed the world's oil supply system is, even in the short term, to unexpected supply disruptions.

2. Recognizes and reaffirms the continuing need for strong policy action in all sectors of energy policy.

3. With regard to coal in particular, notes the study "Steam Coal Prospects to 2000" in which the Secretariat, on the basis of its own analysis, presents views on the potential contribution of coal to the future energy supply and demand balance.

4. Assesses the future potential for coal as follows:

a. Conventional oil supplies will not be available at reasonable prices and in sufficient quantities to meet the growing needs of the world during the medium and longer term.

b. Coal, nuclear power and energy conservation are the principal energy options with major potential for reducing oil dependence and thus improving the overall energy position in the medium term.

c. In many areas of the world, coal is now economically competitive with oil in major energy consumption sectors. The competitiveness of coal is likely to improve in the future. Moreover, the coal resource base is extensive and could sustain a greatly increased production of coal.

d. Long lead times for investment in coal utilization equipment on the demand side and in mining and transportation on the supply side limit the short-term potential to expand coal's contribution. Beyond 1985, coal could provide a substantially greater contribution to the energy needs of IEA countries. This depends on the adoption by governments now of appropriate coal policies which stimulate capital investments on a scale commensurate with the long-term potential of this energy source.

5. Considers that with action along the lines set forth below, thermal coal utilization in the IEA area, which in 1976 was 475 Mtoe, could, by adoption of strong national coal policies, increase beyond the 900 Mtoe for 1990 in current forecasts based on country submissions towards the 1500 Mtoe level for 2000 projected in the Secretariat's accelerated policy case in "Steam Coal Prospects to 2000".

6. Recognizes that the achievement of a rate of steam coal utilization of this order will require long term policies conducive to increased coal utilization, trade, and production in order to encourage the necessary capital investment.

7. Recognizes, however, that countries with different constitutional structures, with different national economic systems, or with differing forms of organization of their coal industries, whether private, public or mixed, will, in their endeavours to increase coal utilization, trade and production, necessarily have to implement and apply these Principles for IEA Action on Coal in ways which are compatible with these situations.

8. Recognizes that some significant coal producers, particularly the United Kingdom and Germany, will wish to maintain policies for safeguarding their domestic coal production at levels required for energy, social or regional policy reasons.

9. Recognizes that increased coal utilization, trade and production must proceed under acceptable environmental conditions. This will require careful planning from the beginning in order to assure a reasonable and continuing balance between energy requirements and environmental requirements.

10. Concludes that long-term oriented, effective and reliable action by Governments of IEA countries is necessary to reduce the uncertainties associated with the coal option and to improve the conditions under which enterprises develop coal utilization, production and trade, including action to ensure the environmental acceptability of coal.

11. Concludes that in order for national coal policy measures of individual IEA countries to reinforce each other and achieve their full potential, their adoption and implementation should be on the basis of international co-operation within the IEA along the following lines:

a. Coal will be established as a major energy source within the IEA group, and countries where coal does not at present contribute significantly to energy balances will consider it as one of the primary alternative energy choices.

b. Countries with the potential for large increases in coal production, in particular Australia, Canada and the United States, will extend their coal production facilities and infrastructure to permit increased domestic use of coal as well as exports consistent with economic and social costs.

c. Other significant coal producing countries, in particular the United Kingdom and Germany, while securing the level of their coal production required by energy, social and regional policies, will accept imported coal rather than oil to meet demand in excess of this level.

d. All IEA countries will enlarge their use of coal; where insufficient coal is available domestically, countries will seek long-term secure supplies of imported coal and provide security of access to markets.

12. Concludes that co-operation with countries which are not members of IEA, and with regional or other groups of countries with significant coal interests, would also make a positive contribution to increased coal use and trade.

13. Concludes that in order to provide reliable long-term conditions for the development of expanded trade and investment in coal, IEA countries should implement domestic measures in a manner which encourages such development.

14. Concludes that it will be necessary to establish procedures for assessment of evolving national coal programmes and policy on an on-going basis, within the IEA. The purpose of such consideration will be to ensure that national coal policy planning in each IEA country proceeds on the basis of co-operation among IEA countries and takes full account of the overall energy situation.

II. Principles for Coal Policy

The Governing Board at Ministerial Level adopts the following Principles for Coal Policy:

15. As an essential element in meeting future energy needs, and in the light of coal's potential as outlined in Section I, IEA countries will, as common objectives, expand:

a. the use of coal as an alternative fuel;
b. the production of coal to meet this increased demand;
c. international trade in coal to meet increased demand.

They will make that commitment clearly known to investors, industry and the general public.

16. They will assess their environmental policies, provisions and practices affecting coal mining, transport and combustion, disposal of coal waste, and land reclamation and, where necessary, will amplify and clarify them, taking into account:

a. technology which is already available and which can be more widely and effectively utilized in commercial applications;

b. the need to develop technologies for additional improvements through research and development, and to commercialize them as soon as they are economically viable;

c. the need to provide long-term reliability for investments by minimizing revisions of environmental standards for existing facilities which disproportionately increase costs in relation to environmental benefits;

d. the need to minimize the complexities, costs and duration of procedures for obtaining necessary approvals.

17. They will assess on a co-operative basis the wider environmental impact of increasing coal production and combustion taking into account the Recommendation on Coal and the Environment adopted by the OECD Council on 8th May, 1979 following a recommendation by the Environment Committee meeting at Ministerial Level on 7th-8th May, 1979.

18. In fulfilling their existing commitment to allow domestic energy prices to reach a level which encourages energy conservation and development of alternative sources of energy, IEA countries will allow coal to develop its full competitive power.

19. They will ensure that the use of oil for electricity generation is minimized by national energy policy planning which, with a minimum of exemptions, precludes new or replacement base load oil-fired capacity; progressively confines oil to middle and peak loads; and makes maximum use of fuels other than oil in dual-fired capacity.

20. They will encourage the development of transportation systems, port facilities and other infrastructure, where necessary, to handle much larger volumes of coal.

21. They will ensure that advanced methods for coal mining, transport and combustion (particularly fluidised bed combustion, improved emissions control and disposal and utilization of solid wastes) and conversion into other fuels, receive high priority in energy R, D & D programmes; and will promote the rapid commercialization of advanced coal technologies which are economically feasible and environmentally acceptable.

22. IEA countries both as producers and consumers will facilitate the expansion of international trade in coal and will do so on a basis which encourages the development of stable relations between consumers and producers, on fair, reasonable and competitive terms, especially by means of long term contracts. They will ensure that an economic, fiscal and investment climate prevails which is conducive to development of coal production, trade and utilization as envisaged in these Principles for IEA Action on Coal.

23. In order to provide reliable conditions for the development of expanded markets for coal, IEA countries which have measures in force concerning international trade in coal, including coal prices, will implement and apply those measures in a manner which fully supports these Principles for IEA Action on Coal; and they will not introduce new measures regarding international trade in coal which are inconsistent with these Principles for IEA Action on Coal except for over-riding reasons concerning the national interest, in which case they will take full account of these Principles for IEA Action on Coal.*

24. Once a long-term contract for international trade in coal has come into force, the governments of IEA countries concerned will not interfere with the implementation of the contract in accordance with its terms, unless they are compelled to do so by severe developments in the coal supply situation occurring in an individual country which threatens that country's wellbeing; or by a severe international energy supply emergency in which latter case they will apply any restrictions on an equitable and non-discriminatory basis.

25. In their efforts to establish an investment climate which encourages the investment necessary to expand coal production, IEA countries will in general maintain positive attitudes towards investment for coal projects, including international investment flows. Insofar as IEA countries have measures in force which provide for review or control of international investment flows, they will implement and apply those measures in a manner which fully supports these Principles for IEA Action on Coal. They will not introduce new measures

* The exception stated at the end of this paragraph is included at the request of Australia and Canada, and accepted by the other IEA countries on the basis of statements to the meeting by these two countries.

regarding international investment flows for coal projects which are inconsistent with these Principles for IEA Action on Coal except for over-riding reasons concerning the national interest, in which case they will take full account of these Principles for IEA Action on Coal.*

26. To help maintain a steady flow of coal in commercial channels on a non-discriminatory basis, IEA countries will monitor the structure and growth of international coal trade as it develops.

27. In designing and implementing their national energy policies, IEA countries will take the specific steps described in the Annex to the Principles for IEA Action on Coal.

* The exception stated at the end of this paragraph is included at the request of Australia and Canada, and accepted by the other IEA countries on the basis of statements to the meeting by these two countries.

ANNEX TO THE PRINCIPLES FOR IEA ACTION ON COAL

Specific Steps to Increase:

Coal Utilization

1. Ensure that national energy policy planning precludes, with a minimum of exceptions, the construction of new or replacement base-load power plants which are exclusively or mainly oil-fired. Exceptions should be permitted only where they are reasonably covered by the following situations:

- national action has been taken to restructure refinery yield patterns toward light products but has not yet been able to eliminate excess quantities of residual fuel oil which cannot be used for other purposes;

- economic or supply conditions, including remoteness of location, are such that use of fuels other than oil is unreasonably expensive in comparison with oil;

- because of local climatic or demographic conditions it is impossible or unreasonably expensive to use fuels other than oil fuels in an environmentally acceptable way even with advanced technology.

2. Require that existing oil-fired base-load power plants be progressively limited to middle or peak load requirements.

3. Ensure that dual-fired power plants are not fired with oil unless other fuels are unreasonably expensive in comparison with oil or it is temporarily necessary for environmental reasons.

4. Facilitate timely construction of coal-fired power plants and supporting facilities where necessary, e.g. by improved siting and licensing procedures.

5. Encourage electric utilities to secure, and coal producers to supply, a larger part of coal requirements for power generation under long-term arrangements in order to achieve long-term supply stability; and facilitate negotiations between electric utilities and coal producers.

6. Encourage the substitution of coal for oil in new and existing industrial facilities for production of steam and process heat, unless the costs, including those for environmental protection, would be unreasonably high in comparison with oil.

7. Encourage the use of large coal-fired boilers when planning new industrial parks, district heating and cogeneration projects.

8. Ensure adequate research and development and facilitate the rapid commercialization of improved technologies for coal combustion, including means for keeping the coal combustion cycle environmentally acceptable.

9. Encourage the commercialization of technologies for converting coal into gas or liquid fuels, including demonstration plants.

Coal Mining

10. Ensure that fiscal regimes, e.g., government royalties and severance taxes, or transportation tariffs do not adversely affect the viability of coal mining developments.

11. Undertake programmes for labour training, improved community infrastructure and other services, where necessary to increase production.

12. Ensure that conditions for leasing of or other access to government lands and for licensing procedures for mine development encourage timely and effective expansion of coal production.

13. Ensure that mining regulations relating to environmental safety and health regulations take account of available technologies.

Coal Transportation

14. Encourage the development of efficient, economic and environmentally acceptable transportation systems with adequate capacity and flexibility (e.g. inland transportation facilities, ocean port facilities and sea-going carriers) where necessary to handle expected increases in coal trade volumes.

Annex II

Decision of the Governing Board on Procedures for Review of IEA Countries' Coal Policies

The Governing Board

Considering that the Governing Board at Ministerial Level on 21st - 22nd May, 1979 adopted the Principles for IEA Action on Coal;

Considering the need for systematic review and assessment of coal policies as well as the need for a framework for consultations among IEA countries on coal policies;

DECIDES:

1. The extent to which the Principles for Coal Policy set forth in Section II of the Principles for IEA Action on Coal and the specific steps described in the Annex thereto are adopted and implemented as part of the national coal policies of individual IEA countries will be a principal element of the IEA process of systematic review.

2. At regular intervals (not ordinarily less than every two years), the Governing Board will assess the prospects for world coal production, trade and utilization, and review the extent to which the main elements of national coal policies required to provide a co-operative framework within the IEA for expansion of coal use and trade have been adopted and implemented. In connection with each such review, IEA countries will inform the Governing Board, inter alia, as to the matters enumerated below, and be prepared to consult with other IEA countries within the Governing Board as to their impact on energy supply and demand:

a. Their trade policy system which secures levels of domestic coal production (paragraph 11c of the Principles);

b. Their proposed level of coal utilisation and in particular the extent to which their national energy policy planning permits the construction of new or replacement base-load power plants which are exclusively or mainly oil-fired, and their reasons therefor (paragraph 1 of Annex to the Principles);

c. Whether their trade policy system and its implementation, insofar as it affects coal, is consistent with the Principles for IEA Action on Coal (paragraphs 22, 23 and 24 of the Principles);

d. The extent to which necessary long-term investments in coal producing and transportation facilities have been forthcoming; whether existing measures in force which provide for review or control of international investment flows have been implemented or applied in a manner which fully supports the Principles for IEA Action on Coal; and whether new measures introduced regarding international investment flows are consistent with the Principles for IEA Action on Coal, (paragraphs 22 and 25 of the Principles).

3. IEA countries will report to the Governing Board and be prepared to consult with other IEA countries under the aegis of the Governing Board, preferably before the event but in any case without delay:

a. Whenever they introduce, or have introduced, a significant change in their policy regarding any of the matters referred to in paragraph 2 of this Decision;

b. As to the nature of any over-riding reasons concerning the national interest which have led to new measures regarding international trade in coal (paragraph 23 of the Principles) or international investment flows for coal projects (paragraph 25 of the Principles), and the measures they intend to apply or have applied;

c. As to the nature of any emergency because of which they intend to interfere, or have interfered, with transactions under long-term coal supply contracts (paragraph 24 of the Principles);

d. Measures envisaged or taken to counter commercial activities which threaten to disrupt the steady flow of coal in commercial channels (paragraph 26 of the Principles).

OECD SALES AGENTS
DÉPOSITAIRES DES PUBLICATIONS DE L'OCDE

ARGENTINA – ARGENTINE
Carlos Hirsch S.R.L., Florida 165, 4° Piso (Galería Guemes)
1333 BUENOS AIRES, Tel. 33.1787.2391 y 30.7122
AUSTRALIA – AUSTRALIE
Australia and New Zealand Book Company Pty, Ltd.,
10 Aquatic Drive, Frenchs Forest, N.S.W. 2086
P.O. Box 459, BROOKVALE, N.S.W. 2100
AUSTRIA – AUTRICHE
OECD Publications and Information Center
4 Simrockstrasse 5300 BONN. Tel. (0228) 21.60.45
Local Agent/Agent local :
Gerold and Co., Graben 31, WIEN 1. Tel. 52.22.35
BELGIUM – BELGIQUE
LCLS
35, avenue de Stalingrad, 1000 BRUXELLES. Tel. 02.512.89.74
BRAZIL – BRÉSIL
Mestre Jou S.A., Rua Guaipa 518,
Caixa Postal 24090, 05089 SAO PAULO 10. Tel. 261.1920
Rua Senador Dantas 19 s/205-6, RIO DE JANEIRO GB.
Tel. 232.07.32
CANADA
Renouf Publishing Company Limited,
2182 St. Catherine Street West,
MONTRÉAL, Que. H3H 1M7. Tel. (514)937.3519
OTTAWA, Ont. K1P 5A6, 61 Sparks Street
DENMARK – DANEMARK
Munksgaard Export and Subscription Service
35, Nørre Søgade
DK 1370 KØBENHAVN K. Tel. +45.1.12.85.70
FINLAND – FINLANDE
Akateeminen Kirjakauppa
Keskuskatu 1, 00100 HELSINKI 10. Tel. 65.11.22
FRANCE
Bureau des Publications de l'OCDE,
2 rue André-Pascal, 75775 PARIS CEDEX 16. Tel. (1) 524.81.67
Principal correspondant :
13602 AIX-EN-PROVENCE : Librairie de l'Université.
Tel. 26.18.08
GERMANY – ALLEMAGNE
OECD Publications and Information Center
4 Simrockstrasse 5300 BONN Tel. (0228) 21.60.45
GREECE – GRÈCE
Librairie Kauffmann, 28 rue du Stade,
ATHÈNES 132. Tel. 322.21.60
HONG-KONG
Government Information Services,
Publications/Sales Section, Baskerville House,
2/F., 22 Ice House Street
ICELAND – ISLANDE
Snaebjörn Jönsson and Co., h.f.,
Hafnarstraeti 4 and 9, P.O.B. 1131, REYKJAVIK.
Tel. 13133/14281/11936
INDIA – INDE
Oxford Book and Stationery Co. :
NEW DELHI-1, Scindia House. Tel. 45896
CALCUTTA 700016, 17 Park Street. Tel. 240832
INDONESIA – INDONÉSIE
PDIN-LIPI, P.O. Box 3065/JKT., JAKARTA, Tel. 583467
IRELAND – IRLANDE
TDC Publishers – Library Suppliers
12 North Frederick Street, DUBLIN 1 Tel. 744835-749677
ITALY – ITALIE
Libreria Commissionaria Sansoni :
Via Lamarmora 45, 50121 FIRENZE. Tel. 579751
Via Bartolini 29, 20155 MILANO. Tel. 365083
Sub-depositari :
Editrice e Libreria Herder,
Piazza Montecitorio 120, 00 186 ROMA. Tel. 6794628
Libreria Hoepli, Via Hoepli 5, 20121 MILANO. Tel. 865446
Libreria Lattes, Via Garibaldi 3, 10122 TORINO. Tel. 519274
La diffusione delle edizioni OCSE è inoltre assicurata dalle migliori librerie nelle città più importanti.
JAPAN – JAPON
OECD Publications and Information Center,
Landic Akasaka Bldg., 2-3-4 Akasaka,
Minato-ku, TOKYO 107 Tel. 586.2016
KOREA – CORÉE
Pan Korea Book Corporation,
P.O. Box n° 101 Kwangwhamun, SÉOUL. Tel. 72.7369
LEBANON – LIBAN
Documenta Scientifica/Redico,
Edison Building, Bliss Street, P.O. Box 5641, BEIRUT.
Tel. 354429 – 344425
MALAYSIA – MALAISIE
and/et **SINGAPORE – SINGAPOUR**
University of Malaysia Co-operative Bookshop Ltd.
P.O. Box 1127, Jalan Pantai Baru
KUALA LUMPUR. Tel. 51425, 54058, 54361
THE NETHERLANDS – PAYS-BAS
Staatsuitgeverij
Verzendboekhandel Chr. Plantijnstraat 1
Postbus 20014
2500 EA S-GRAVENAGE. Tel. nr. 070.789911
Voor bestellingen: Tel. 070.789208
NEW ZEALAND – NOUVELLE-ZÉLANDE
Publications Section,
Government Printing Office Bookshops:
AUCKLAND: Retail Bookshop: 25 Rutland Street,
Mail Orders: 85 Beach Road, Private Bag C.P.O.
HAMILTON: Retail Ward Street,
Mail Orders, P.O. Box 857
WELLINGTON: Retail: Mulgrave Street (Head Office),
Cubacade World Trade Centre
Mail Orders: Private Bag
CHRISTCHURCH: Retail: 159 Hereford Street,
Mail Orders: Private Bag
DUNEDIN: Retail: Princes Street
Mail Order: P.O. Box 1104
NORWAY – NORVÈGE
J.G. TANUM A/S Karl Johansgate 43
P.O. Box 1177 Sentrum OSLO 1. Tel. (02) 80.12.60
PAKISTAN
Mirza Book Agency, 65 Shahrah Quaid-E-Azam, LAHORE 3.
Tel. 66839
PHILIPPINES
National Book Store, Inc.
Library Services Division, P.O. Box 1934, MANILA.
Tel. Nos. 49.43.06 to 09, 40.53.45, 49.45.12
PORTUGAL
Livraria Portugal, Rua do Carmo 70-74,
1117 LISBOA CODEX. Tel. 360582/3
SPAIN – ESPAGNE
Mundi-Prensa Libros, S.A.
Castelló 37, Apartado 1223, MADRID-1. Tel. 275.46.55
Libreria Bosch, Ronda Universidad 11, BARCELONA 7.
Tel. 317.53.08, 317.53.58
SWEDEN – SUÈDE
AB CE Fritzes Kungl Hovbokhandel,
Box 16 356, S 103 27 STH, Regeringsgatan 12,
DS STOCKHOLM. Tel. 08/23.89.00
SWITZERLAND – SUISSE
OECD Publications and Information Center
4 Simrockstrasse 5300 BONN. Tel. (0228) 21.60.45
Local Agents/Agents locaux
Librairie Payot, 6 rue Grenus, 1211 GENÈVE 11. Tel. 022.31.89.50
Freihofer A.G., Weinbergstr. 109, CH-8006 ZÜRICH.
Tel. 01.3634282
TAIWAN
Good Faith Worldwide Int'l Co., Ltd.
9th floor, No. 118, Sec. 2
Chung Hsiao E. Road
TAIPEI. Tel. 391.7396/391.7397
THAILAND – THAILANDE
Suksit Siam Co., Ltd., 1715 Rama IV Rd,
Samyan, BANGKOK 5. Tel. 2511630
TURKEY – TURQUIE
Kültur Yayinlari Is-Türk Ltd. Sti.
Atatürk Bulvari No : 77/B
KIZILAY/ANKARA. Tel. 17 02 66
Dolmabahce Cad. No : 29
BESIKTAS/ISTANBUL. Tel. 60 71 88
UNITED KINGDOM – ROYAUME-UNI
H.M. Stationery Office, P.O.B. 569,
LONDON SE1 9NH. Tel. 01.928.6977, Ext. 410 or
49 High Holborn, LONDON WC1V 6 HB (personal callers)
Branches at: EDINBURGH, BIRMINGHAM, BRISTOL,
MANCHESTER, CARDIFF, BELFAST.
UNITED STATES OF AMERICA – ÉTATS-UNIS
OECD Publications and Information Center, Suite 1207,
1750 Pennsylvania Ave., N.W. WASHINGTON, D.C.20006 – 4582
Tel. (202) 724.1857
VENEZUELA
Libreria del Este, Avda. F. Miranda 52, Edificio Galipan,
CARACAS 106. Tel. 32.23.01/33.26.04/33.24.73
YUGOSLAVIA – YOUGOSLAVIE
Jugoslovenska Knjiga, Terazije 27, P.O.B. 36, BEOGRAD.
Tel. 621.992

Les commandes provenant de pays où l'OCDE n'a pas encore désigné de dépositaire peuvent être adressées à :
OCDE, Bureau des Publications, 2, rue André-Pascal, 75775 PARIS CEDEX 16.

Orders and inquiries from countries where sales agents have not yet been appointed may be sent to:
OECD, Publications Office, 2 rue André-Pascal, 75775 PARIS CEDEX 16.

OECD PUBLICATIONS, 2, rue André-Pascal, 75775 PARIS CEDEX 16 - No. 42259 1982
PRINTED IN FRANCE
(61 82 07 1) ISBN 92-64-12336-9